Springer Theses

Recognizing Outstanding Ph.D. Research

Aims and Scope

The series "Springer Theses" brings together a selection of the very best Ph.D. theses from around the world and across the physical sciences. Nominated and endorsed by two recognized specialists, each published volume has been selected for its scientific excellence and the high impact of its contents for the pertinent field of research. For greater accessibility to non-specialists, the published versions include an extended introduction, as well as a foreword by the student's supervisor explaining the special relevance of the work for the field. As a whole, the series will provide a valuable resource both for newcomers to the research fields described, and for other scientists seeking detailed background information on special questions. Finally, it provides an accredited documentation of the valuable contributions made by today's younger generation of scientists.

Theses are accepted into the series by invited nomination only and must fulfill all of the following criteria

- They must be written in good English.
- The topic should fall within the confines of Chemistry, Physics, Earth Sciences, Engineering and related interdisciplinary fields such as Materials, Nanoscience, Chemical Engineering, Complex Systems and Biophysics.
- The work reported in the thesis must represent a significant scientific advance.
- If the thesis includes previously published material, permission to reproduce this must be gained from the respective copyright holder.
- They must have been examined and passed during the 12 months prior to nomination.
- Each thesis should include a foreword by the supervisor outlining the significance of its content.
- The theses should have a clearly defined structure including an introduction accessible to scientists not expert in that particular field.

More information about this series at http://www.springer.com/series/8790

Wai Hou (Alan) Lio

Blade-Pitch Control for Wind Turbine Load Reductions

Doctoral Thesis accepted by
the University of Sheffield, Sheffield, UK

Springer

Author
Dr. Wai Hou (Alan) Lio
Department of Wind Energy
Technical University of Denmark
Roskilde
Denmark

Supervisors
Dr. J. A. Rossiter
The University of Sheffield
Sheffield
UK

Dr. B. Ll. Jones
The University of Sheffield
Sheffield
UK

ISSN 2190-5053 ISSN 2190-5061 (electronic)
Springer Theses
ISBN 978-3-030-09257-3 ISBN 978-3-319-75532-8 (eBook)
https://doi.org/10.1007/978-3-319-75532-8

Printed on acid-free paper

This Springer imprint is published by the registered company Springer International Publishing AG part of Springer Nature
The registered company address is: Gewerbestrasse 11, 6330 Cham, Switzerland

Dedicated to my family

Supervisors' Foreword

The potential, or one might even say, essential benefits to the world of green energy, are becoming ever more evident. Of the various production techniques, wind energy is perhaps the most widely adopted, and indeed, there is an increasing recognition that wind energy can be produced at prices comparable to traditional energy sources, especially if one is able to increase the turbine blade dimensions and site turbines in suitable locations, such as out at sea. In order to make these new turbines economic, a number of new challenges are being presented, one of which is to manage the fatigue life of the core components. A turbine is only economic if operating for a large number of years. Any faults or failures requiring intervention are very expensive to fix and seriously impact on the long-term economic viability.

Wai Hou Lio's work tackles some of these issues head-on and in particular focuses on the fatigue life of the blades and connected components. Fatigue life is linked to peak loads and load variation on the blades, and thus, a core aim is to reduce these loads while ensuring the turbine continues to produce the required power output. A key thrust here is to understand the origin of the moments and forces on the blades and exploit this knowledge with a suitably tailored control strategy.

Some of the ideas Wai Hou Lio has proposed, such as a mechanism which allows a convenient and useful separation of feed-forward and feedback modes within predictive control, are novel in the field and backed up by new theory he developed. This is particularly applicable to strategies for minimising load variations on turbine blades with the knowledge of the upcoming wind. The work on 'Fundamental performance similarities between individual pitch control strategies for wind turbines' establishes a rigorous mathematical proof that answers a long-standing problem in the area of wind turbine control related to the best way of individually controlling the blades of a wind turbine to minimise harmful fatigue loads caused by uneven and unsteady forcing from the oncoming wind. By solving this problem, he has essentially unified a research area, and his contribution now provides wind turbine control engineers greater clarity in designing blade pitch control systems for the next generation of multi-MW offshore wind turbines. He has been imaginative in considering which problems may be amenable to a different approach, one example of this being the thoughts on estimating tower motion from blade sensors, thus either increasing redundancy or reducing cost.

We are extremely pleased that Wai Hou Lio's significant, excellent and timely research work has been selected for publication in the prestigious Springer Theses series.

Sheffield, UK
October 2017

Dr. J. A. Rossiter
Dr. B. Ll. Jones

Abstract

Large wind turbines are subjected to the harmful loads that arise from the spatially uneven and temporally unsteady oncoming wind. Such loads are the known sources of fatigue damage that reduce the turbine operational lifetime, ultimately increasing the cost of wind energy to the end-users. In recent years, a substantial amount of studies has focused on blade pitch control and the use of real-time wind measurements, with the aim of attenuating the structural loads on the turbine blades and rotor.

However, many of the research challenges still remain unsolved. For example, there are many classes of blade individual pitch control (IPC) techniques, but the link between these different but competing IPC strategies was not well investigated. In addition, another example is that many studies employed model predictive control (MPC) for its capability to handle the constraints of the blade pitch actuators and the measurement of the approaching wind, but often, wind turbine control design specifications are provided in frequency domain that is not well taken into account by the standard MPC.

To address the missing links in various classes of the IPCs, this thesis aims to investigate and understand the similarities and differences between each of their performance. The results suggest that the choice of IPC designs rests largely with preferences and implementation simplicity. Based on these insights, a particular class of the IPCs lends itself readily for extracting tower motion from measurements of the blade loads. Thus, this thesis further proposes a tower load reduction control strategy based solely upon the blade load sensors.

To tackle the problem of MPC on wind turbines, this thesis presents an MPC layer design upon a predetermined robust output-feedback controller. The MPC layer handles purely the feed-forward and constraint knowledge, whilst retaining the nominal robustness and frequency-domain properties of the predetermined closed loop. Thus, from an industrial perspective, the separate nature of the proposed control structure offers many immediate benefits. Firstly, the MPC control can be implemented without replacing the existing feedback controller. Furthermore, it provides a clear framework to quantify the benefits in the use of advance real-time measurements over the nominal output-feedback strategy.

Parts of this thesis have been published in the following publications:

Journal articles

[1] **Lio WH**, Jones BL & Rossiter JA (2017) Preview predictive control layer design based upon known wind turbine blade-pitch controllers. Wind Energy, 20:1207–1226.
[2] **Lio WH**, Jones BL & Rossiter JA (2017) Fundamental performance similarities between individual pitch control strategies for wind turbines, International Journal of Control, 90(1):37–52.

Peer-reviewed conference articles

[1] **Lio WH**, Jones BL & Rossiter JA (2016) Analysis and design of a tower motion estimator for wind turbines, 2016 International Conference on Renewable Energy Research and Applications (ICRERA), Birmingham, UK, 20–23 Nov 2016.
[2] **Lio WH**, Rossiter JA & Jones BL (2016) Predictive control design on an embedded robust output-feedback compensator for wind turbine blade-pitch preview control, 2016 European Control Conference (ECC), Aalborg, Denmark, 29 Jun–1 Jul 2016.
[3] **Lio WH**, Rossiter JA & Jones BL (2014) A review on applications of model predictive control to wind turbines, 2014 UKACC International Conference on Control (CONTROL), Loughborough, UK, 8–11 Jul 2014.

Acknowledgements

I would like to express my sincere gratitude to all the people that made this thesis possible.

First and foremost, to my Ph.D. supervisors Dr. Anthony Rossiter and Dr. Bryn Jones, for their continuous support and kind supervision throughout my Ph.D. journey. I am incredibly lucky to have two inspirational supervisors, who both shared with me their invaluable experience and offered me encouragement whenever I encountered challenges. Thank you both so much.

To Prof. Ron Patton and Prof. Zi-Qiang Lang, for the insightful advice and thoughts to my thesis. I also would like to express my gratitude to Dr. Eric Kerrigan, Dr. Bing Feng Ng, Dr. Stefano Longo, Dr. Ian Couchman and Dr. Chris Spruce, who introduced me to the interesting topics of control and wind energy.

To my colleagues and friends in Sheffield, especially Dr. Evans Ejegi, Dr. Adham Alsharkawi, Dr. Shukri Dughman and Dr. Yahya Al-naumani, for their friendships and supports. In particular, I would like to thank Dr. Qian Lu for helping with the simulation codes in the early stage of my Ph.D.

To my family, my mother In Fong Leong and my sister Dr. Ka Ian Lio, for their lifelong love, support and motivation. Last but not least, I am truly indebted to my dearest wife, Wanjing Zhao, for her altruistic care, love and encouragement.

Roskilde, Denmark
January 2018

Dr. Wai Hou (Alan) Lio

Contents

Nomenclature

$\mathbb{R}$	Set of real numbers
$\mathbb{C}$	Set of complex numbers
$\mathbb{Z}$	Set of integers
$\mathcal{R}$	Space of proper real-rational transfer function matrices
t	Time
k	Time step index
f	Frequency
s	Complex variable
$.^{\mathrm{T}}$	Matrix conjugate transpose
$.^{-1}$	Matrix inverse
$\dot{x} := \frac{dx}{dt}$	Differentiation of x
$j := \sqrt{-1}$	Imaginary unit
I	Identity matrix
$\Vert\cdot\Vert$	Vector norm
$P^*(s) := P(-s)^{\mathrm{T}}$	Adjoint of $P(s) \in \mathcal{R}$

Wind Turbines

θ	Pitch angle of blades
υ	Stream-wise wind speed
ρ	Air density
$\omega_{\mathrm{t}}, \omega_{\mathrm{b}}$	Natural frequencies of tower and blades
$\zeta_{\mathrm{t}}, \zeta_{\mathrm{b}}$	Damping ratios of the tower and blades
ϕ	Blade azimuth angle
ω	Rotational speed of rotor
M	Flap-wise blade root bending moment
$\dot{x}_{\mathrm{fa}}$	Fore-and-aft translational velocity of tower-top
$\dot{\varphi}_{\mathrm{fa}}$	Fore-and-aft rotational velocity of tower-top

Control Systems

A, B, C, D	State-space matrices
x, u, y, d	Vectors of system state, input, output and disturbance
c	Perturbation to the feedback law
$n_{\mathrm{x}}, n_{\mathrm{u}}, n_{\mathrm{y}}, n_{\mathrm{d}}$	Dimensions of systems state, input, output and disturbance
n_{c}	Control horizon
n_{a}	Preview horizon
$\hat{x}$	Estimated value of x
J	Cost
b	Robust stability margin
$x_{i\|k}$	Prediction of x at sample time $k+i$ given the information available at sample time k
$\underrightarrow{x}_k$	Vector of prediction of x at time k

List of Figures

List of Tables

Chapter 1
Introduction

1.1 Motivation

Wind energy is a sustainable and environmentally friendly source of energy. To tackle the problems of climate change and global warming, there has been a rapid growth in utilising the wind power as a source of electricity generation in many countries. Amongst the total installed capacity, an increasing number of large wind turbines begins installing offshore for exploiting the abundant offshore wind resources (Ahlgrimm, 2016). Nonetheless, reliable operation of wind turbines is the key factor for cost-effective offshore wind energy harvesting and particularly, the maintenance and repair activities become costly for offshore wind turbines located in a remote area. Thus, technologies that can reduce the costly maintenance and repair activities are of significant importance. One of these is the use of blade pitch control as a means of attenuating the harmful turbine structural loads that arise from spatially uneven and temporarily unsteady oncoming wind. Load mitigations on the key turbine components not only improve the reliability of the operation, but also lead to a significant reduction in the cost of the required materials. Therefore, this research is motivated to investigate the blade-pitch control strategies for wind turbine load reductions.

1.2 Overview of Blade-Pitch Control

1.2.1 Blade-Pitch Control

Figure 1.1 depicts a typical three-bladed horizontal axis wind turbine. The main components of a typical wind turbine are blades, hub, nacelle and tower. In a modern megawatt turbine, there are numerous actuators and sensors available for control and monitoring purposes. Thus, a brief introduction of the actuators and sensors is provided. Inside the turbine blades, pitch actuators, that are typically hydraulic or

W. H. (Alan) Lio, *Blade-Pitch Control for Wind Turbine Load Reductions*, Springer Theses, https://doi.org/10.1007/978-3-319-75532-8_1

Fig. 1.1 A typical three-bladed variable-speed horizontal-axis wind turbine

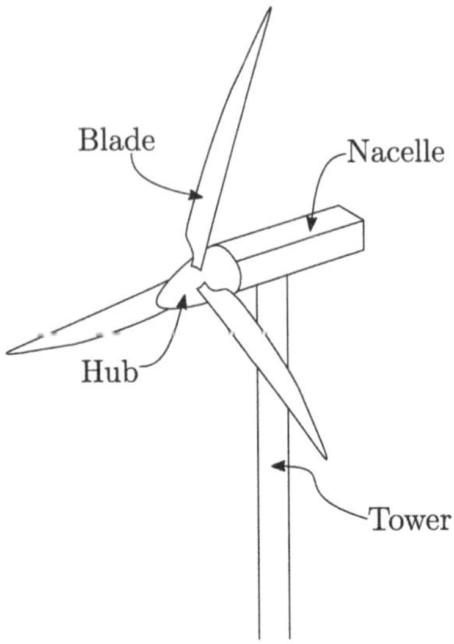

electric motors, are used to adjust the pitch angles of the blades. In addition, blade load sensors, typically strain gauges or optical fibres, are employed at the root of the blades in order to to obtain measurements of the flap-wise or edge-wise blade root bending moments. Generator and drive train are located inside the nacelle where generator and rotor speed sensors are used in order to measure the rotational speed. To obtain the wind speed measurements, an anemometer is employed at the top of the nacelle to monitor the hub-height wind speed. In recent years, studies began exploiting remote wind sensing technology, such as light and detection ranging (LIDAR) systems, to measure the real-time upcoming wind information.

Figure 1.2 describes the working principle of a typical modern wind turbine. As the air flow passes the surfaces of the blades, a pair of aerodynamic forces, namely lift and drag, are generated. The lift and drag are dependent on the aerofoil of the blade and pitch angle of the blades θ. These forces can be expressed as an in-plane force, that causes the blade to rotate, and an out-of-plane force, that induces the nacelle and tower fore-aft motions. As the rotor spins, the rotational energy of the rotor is then transmitted via the drive trains to the generator, where the energy is converted into the electrical form. Notice that the in-plane and out-of-plane forces on the blade can be manipulated by the blade pitch angle. As a result, the rotational speed of the rotor and the motions of the blade and tower can be controlled by changing the blade pitch angles.

Over the past few decades, blade pitch control systems have been widely used for the purposes of regulating the rotational speed of the rotor. In a situation where the wind speed is higher than a specified rated value, the turbine would produce a power output that is higher than the nominal power if blade pitch control systems are not employed. Notice that these operating conditions are known as the above-rated wind conditions. In such conditions, any excessive amount of the aerodynamic torques

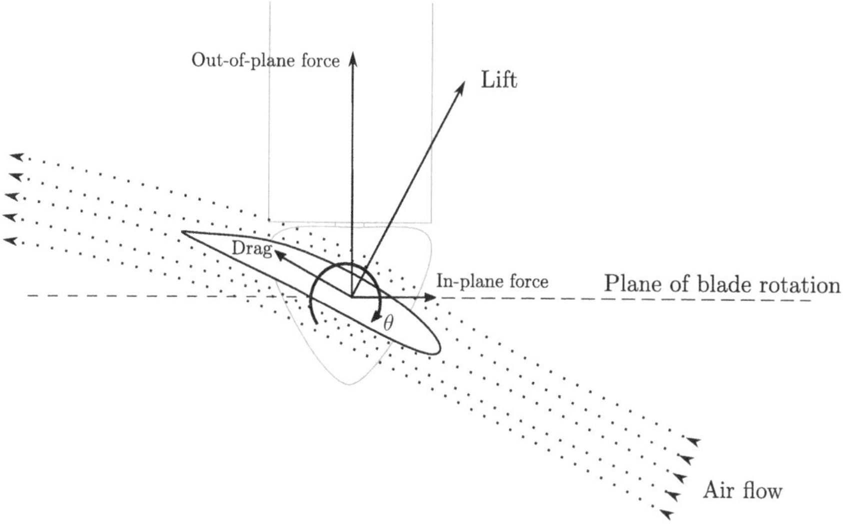

Fig. 1.2 A brief walk-through of the working principle of a typical horizontal-axis wind turbine

on the blades will cause the rotor to spin faster than its rated speed, resulting in damage to the mechanical components and generator. Thus, blade pitch controllers are critical for alleviating any excessive aerodynamic torques on the blades, where the pitch angle of each blade is collectively adjusted by the same amount in response to the rotor speed measurement. This control strategy is well known as the blade collective pitch control (CPC).

In recent years, an increasing amount of studies investigated the possibility of utilising the blade pitch control system for attenuating unsteady loads on the turbine blade, rotor and tower structures. Such loadings arise from a number of sources, such as wind shear, tower shadow, yaw misalignment and turbulence. For example, the wind shear causes the blade to experience an uneven force at every blade rotation. Since the use of blade pitching can alter the aerodynamic forces on the blades, such loadings caused by the wind shear can be alleviated by adjusting the pitch angle of each blade independently. This technique is known as blade individual pitch control (IPC). The IPC provides an additional blade pitch demand to the control actions from the CPC, typically in response to the measurements of the flap-wise blade root bending moments. Notice that the CPC regulates the rotor at frequencies below the rotational frequency of the rotor, whilst the IPC attenuates blade load at the harmonics of the blade rotational frequencies. Given the frequency range of the targeted loads is different for each controller, thus, separate loop design is prevalent in the industry and academia (Bossanyi, 2003; van Engelen, van der Hooft, & Schaak, 2001). Therefore, it is significantly important to ensure there is no coupling between the CPC and IPC loop when designing the controllers. More details are explained in Sect. 2.8 in Chap. 2.

1.2.2 *Preview and Model Predictive Control*

In pursuit of effective load reductions or rotor speed regulations by blade pitching, the pitch actuators are sometimes utilised to their limits. Consequently, this motivates the use of model predictive control (MPC) for blade pitch control strategies. Typically, an MPC control algorithm makes predictions over a finite horizon based on a mathematical model of the plant, past information of the inputs and outputs and sometimes future measurements. Subsequently, the algorithm optimises a future input sequence subject to the predictions and constraint knowledge. The first calculated input is then applied to the plant and the entire process repeats at every sample time. In a typical wind turbines, there are physical limitations on the blade pitch actuators as well as some operational requirements such as the maximal limit on the rotor speed.

In recent years, the development of remote sensing technologies, such as light detection and ranging (LIDAR) systems, enables accurate measurements of the approaching wind, which allows anticipatory blade pitch adjustments to counteract the advance disturbance via feed-forward control design. Thus, this is another main reason for motivating the use of MPC since the preview measurements could be incorporated into the predictive control law systematically (e.g. Laks, 2013). Based on the preview wind measurements, MPC controllers can plan the control moves that compensate the upcoming wind disturbance within the feasible range of the blade pitch actuators, resulting in better load reductions or rotor speed regulation for wind turbines. More details regarding model predictive control are discussed in Sect. 2.6.

1.3 Challenges in Blade-Pitch Control and Aims of the Thesis

The blade collective pitch control (CPC) strategies have been studied widely in the academic community and employed in most of the modern variable-speed wind turbines for the purpose of rotor speed regulation. In contrast, the concepts of blade individual pitch control (IPC), that enable load attenuations on the blades and rotor, started attracting more attentions from the academia as well as the industry in recent years. This is driven by the need for cost-competitive wind turbines. Given more choices of actuators and sensors available to modern wind turbines (e.g. LIDAR systems) or the changes in operating locations from onshore to offshore, there are still many challenges and opportunities regarding the blade pitch control technology. Some specific research questions addressed by this thesis are listed as follows.

1.3.1 Challenges in Blade-Pitch Control

Clarifications amongst various classes of IPC strategies

One of the earliest studies in individual pitch control (IPC) dated back to 2003 by Bossanyi (2003). Since then, there are hundreds of published works related to IPC strategies for alleviating unsteady loads on the blades and rotor. Of the many IPC strategies that have been published in recent years, most can be grouped into two distinct classes, characterised by the specific turbine loads that they are primarily designed to attenuate. The first and most popular branch is to target turbine loads upon fixed and non-rotating turbine structures, such as the rotor, tower, main shaft bearing. The second branch is to attenuate loads upon the rotating turbine structures, primarily the blades. Many types of coordinate transformations have been employed in these two classes of IPC strategies, for example, Coleman (Bossanyi, 2003) and Clarke transformations (Zhang, Cheng, & Chen, 2013). Also, various control designs are proposed based on the transformed blade models, for example, single-input-single-output (SISO) or multi-input-multi-output (MIMO) techniques. The main question is whether there are any similarities between these various designs and under what conditions that their equivalence is valid. If the performance similarities could be established, these promises enable a simple SISO design to perform as well as a complicated MIMO design for the IPC strategies. A thorough review of the existing IPC strategies is provided in Sect. 3.2 in Chap. 3.

Potential usage of blade load sensors to estimate tower motions

Typically, the use of blade individual pitch control strategies requires installing and commissioning of blade load sensors. The same story holds for tower damping control that relies upon measurements from tower-top accelerometers. Nonetheless, there exist strong interactions between the blades and turbine tower and these couplings make estimations of the fore-aft tower-top motion possible based solely upon the measurement of the blade sensors. However, the measurement of the blade moments are generated in a rotating frame, whilst the tower-top motions are upon a fixed reference frame, thus, modelling such dynamics yields a linear time-periodic system. The question is how to construct a linear model that simplifies the blade and tower dynamics and is suitable for a simple estimator design. Also, how reliable are the estimated tower signals for control purposes?

Systematic design to integrate the preview and constraint knowledge into an existing feedback controller

The upcoming disturbance information is often available for feed-forward design in some control applications. Typically, model predictive control (MPC) is one of the favourable candidates for its capability to handle constraints and preview information systematically. As mentioned earlier, the turbine blade loads exist mainly at the harmonics of the blade rotational frequency, and such performance specifications are often given in the frequency-domain, thus, it is more trivial to synthesise a robust feedback controller using frequency-domain techniques than the standard time-domain MPC approach. Therefore, it is beneficial if the constraint and feed-forward knowledge can be incorporated into a pre-determined control law, that is designed by frequency-domain methods or other methods favoured by the industry. Nonetheless, the standard MPC design based on an existing closed-loop could introduce an extra feedback loop to the plant, where the optimisation couples the feedback and feed-forward information, resulting in a possible deterioration of the robustness and frequency-domain properties[1] of the predetermined closed-loop. As a consequence, the main challenge is to design a modular MPC layer based on an existing feedback control law and most importantly, this MPC layer needs to handle purely the feed-forward and constraint knowledge, without affecting the frequency-domain properties of the pre-determined controller.

Clear case studies of utilising knowledge of the approaching wind and constraints in wind turbines

In recent years, the advancement in remote wind sensing technology enable the availability of cheap and reliable real-time upcoming wind measurements to wind turbines. In addition, in pursuit of effective load attenuations with preview knowledge, the pitch actuators might operate near their limitations. Many studies employed the standard MPC approach for its constraint and preview measurement handling capabilities, where the optimisation takes account of the feedback and feed-forward information, resulting in one single integrated predictive control law that handles the feedback and preview knowledge together. Subsequently, the performance comparison of this predictive control law is often made against a baseline feedback controller. Thus, the performance gained by utilising preview and constraint information could not be distinguished from the feedback control law. As a consequence, from an industrial perspective, a clear and transparent comparison is beneficial that reveals the true benefit of uses of remote wind sensing devices and constraint handling capabilities.

1.3.2 Aims of the Thesis

The main aims of this thesis are to investigate load reduction control strategies for wind turbines via individual blade pitching and use of upcoming wind measurements, with a view to providing systematic analysis and design guidelines on blade pitch and predictive control design in wind turbines. This thesis is divided into two parts, where the first part, in Chaps. 4 and 5, is regarding the individual pitch control design, whilst the latter part, in Chaps. 6 and 7, focuses on feed-forward model predictive control design of wind turbines. The main objectives of this research are outlined as follows:

---

[1]The robustness property in this thesis implies the robust stability margin, that is a generalisation of the gain and phase margins for multi-input-multi-output systems. Details are discussed in 2.8.2.1.

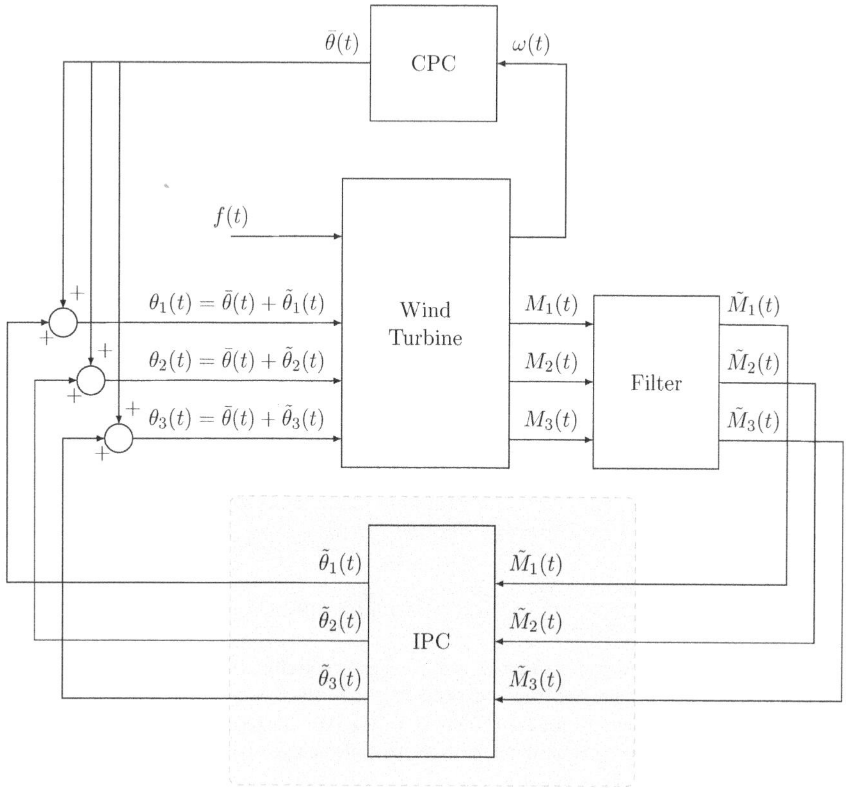

Fig. 1.3 System architecture of a wind turbine employed in Chap. 4, combining collective pitch control (CPC) and individual pitch control (IPC). The CPC regulates rotor speed $\omega(t)$ while the IPC (shaded) attenuates perturbations in the flap-wise root bending moments on each blade $\tilde{M}_{1,2,3}(t)$. Additional inputs to the turbine such as wind loading and generator torque are accounted for in the term $f(t)$

- Investigation of the links between various classes of IPC designs and demonstration of their similarities via analytical and stimulation studies. For example, derivations of the robust stability margin of these different classes of IPCs. And illustration of the numerical results highlighting the load attenuation performance of each IPC. Figure 1.3 shows a typical system architecture of a wind turbine with CPC and IPC.
- Study of the interactions between the turbine blades and tower. Development of an estimation design, that uses measurements from blade load sensors to reconstruct the tower vibration signals, and a control algorithm, that attenuates tower loads based on the estimated signals. Figure 1.4 shows the proposed tower estimator and controller.

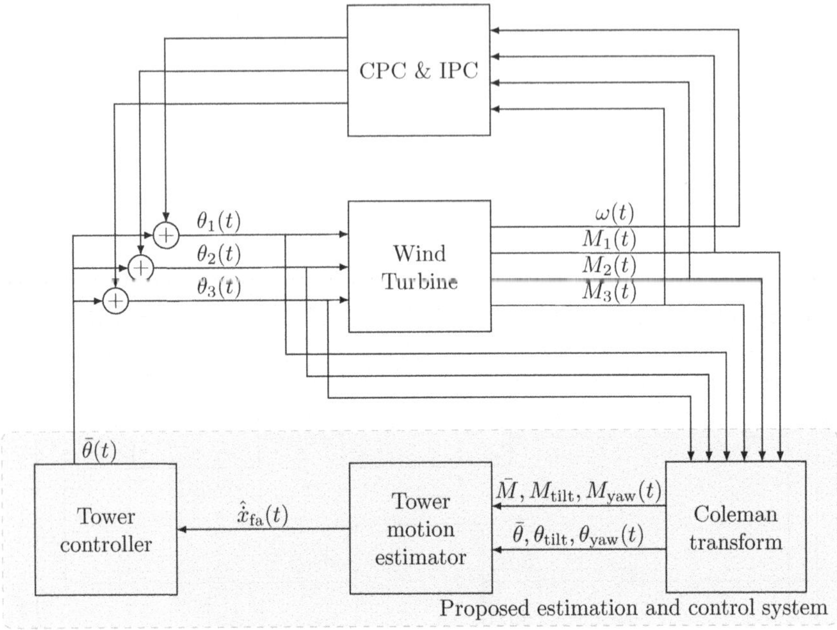

Fig. 1.4 Schematic of the proposed tower estimator and controller. Based upon the pitch input $\theta_{1,2,3}(t)$ and measurements of the blade moments $M_{1,2,3}(t)$, the tower motion estimator with the Coleman transform reconstructs the tower velocity $\hat{\dot{x}}_{\text{fa}}(t)$. Also, the estimated signal can be used for attenuating tower loads by controlling the collective pitch angle $\bar{\theta}(t)$

- Analysis of the standard MPC design that shows how the optimisation couples the feed-forward information with the feedback. Development of a novel modular MPC algorithm based on an existing feedback controller where the MPC layer utilises only the feed-forward measurements and constraint handling whilst retaining the robustness and frequency-domain properties properties of the latter. Figure 1.5 depicts the MPC layer control architecture.
- Study of issues which arise from implementation of the MPC layer in the non-linear simulation wind turbine, for example, the feasibility issues. Illustration of numerical studies, that investigates the constraint handling scenarios. In addition, examination of the performance gained by utilising the information of the upcoming wind and constraints compared to the baseline feedback controller.

The controller and estimator design in this thesis are all validated upon a widely-used high fidelity simulation wind turbine, NREL 5MW baseline turbine (Jonkman & Buhl Jr., 2005; Jonkman, Butterfield, Musial, & Scott, 2009).

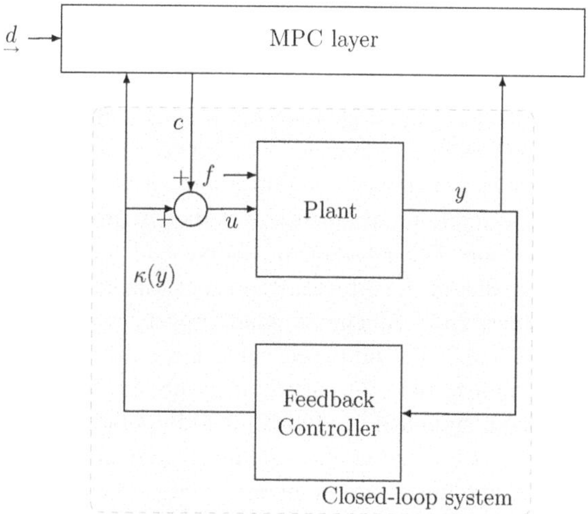

Fig. 1.5 Schematic of model predictive control layer on top of an existing feedback controller. Based upon the controller input $\kappa(y)$, plant output y, the layer optimises the perturbation c that can handle constraints and act upon feed-forward information $\underset{\rightarrow}{d}$. The notation f denotes the disturbance to the plant

1.4 Thesis Layout

This thesis consists of eight chapters, which are structured as follows:

In Chap. 2, an introduction to modern wind turbine operational modes and control objectives are presented at the beginning. In addition, backgrounds on modelling of wind turbines are provided. This is then followed by a brief introduction of model predictive control. Subsequently, methods of fatigue load assessment are discussed. Finally, the details of the simulation package and design of the baseline controller are presented.

In Chap. 3, a review of state-of-the-art individual pitch control strategies is presented. This is then followed by a recent development of feed-forward model predictive control designs. Lastly, recent history and development of feed-forward control in wind turbines is discussed.

In Chap. 4, the similarities of various classes of individual pitch control methods are investigated. The links between these IPC strategies are analysed and established, and furthermore, the key proof of their equivalence in terms of robust stability margin is also derived. Finally, it is then demonstrated that the performance of these IPC techniques is fundamentally equivalent by conducting simulations on a high-fidelity turbine.

In Chap. 5, the interactions between the turbine tower and blades are investigated. Based on these couplings, a linear time-periodic model is yielded and an observability analysis is conducted on such a time-periodic model. Subsequently, coordinate transformation is introduced that simplify the linear time-periodic model into one that is time-invariant. Thus, a linear estimator is then proposed that can reconstruct the tower fore-aft velocity based on the measurements of the flap-wise blade root moments, which are already accessible to the IPC strategies. Finally, a control algorithm that uses the reconstructed tower signals is designed for purposes of tower load attenuations. Simulation examples are provided to show the performance of the estimator and the estimation-based control algorithm.

In Chap. 6, a model predictive control layer is proposed that can systematically integrate the upcoming wind information and constraint handling capability into an existing feedback controller. The first part of the chapter shows a counter-example how the standard MPC optimisation based on a given output-feedback controller can deteriorate the robustness or other frequency properties of the pre-determined closed-loop. The subsequent part presents a novel modular MPC layer design that only handles the feed-forward and constraint information whilst retaining the nominal robustness and frequency-domain properties of the pre-determined closed-loop.

In Chap. 7, the proposed MPC layer design is implemented on a high-fidelity simulation turbine. In particular, issues such as feasibility are studied and a soft constraint approach is adopted to address these issues. To verify the benefit gained of utilising the upcoming wind measurements and constraint handling capabilities, three simulation cases are investigated with constraints imposed on the rotor speed, magnitude and rate of the pitch actuators. Furthermore, sufficiently long simulations were conducted to demonstrate the performance of the proposed control structure.

Finally, in Chap. 8, the contributions of the thesis are summarised, and final conclusions are also presented. At the end, potential avenues for future work are discussed.

References

Ahlgrimm, J. (2016). *IEA wind 2015 annual report* (Tech. Rep.). International Energy Agency.

Bossanyi, E. A. (2003). Individual blade pitch control for load reduction. *Wind Energy*, *6*(2), 119–128.

Jonkman, J., & Buhl Jr, M. (2005). *FAST user's guide* (Tech. Rep.). National Renewable Energy Laboratory (NREL).

Jonkman, J., Butterfield, S., Musial, W., & Scott, G. (2009). *Definition of a 5-MW reference wind turbine for offshore system development* (Tech. Rep.). Golden, CO: National Renewable Energy Laboratory (NREL).

Laks, J. H. (2013). *Preview scheduled model predictive control for horizontal axis wind turbines* (Ph.D. thesis). University of Colorado.

van Engelen, T. G., van der Hooft, E. L., & Schaak, P. (2001). Development of wind turbine control algorithms for industrial use. In *European Wind Energy Conference* (pp. 1098–1101).

Zhang, Y., Cheng, M., & Chen, Z. (2013). Proportional resonant individual pitch control for mitigation of wind turbines loads. *IET Renewable Power Generation*, *7*(3), 191–200.

Chapter 2
Background of Wind Turbine Blade-Pitch Load Reduction Control

2.1 Overview

The aim of this chapter is to provide some background regarding to the wind turbine operations, modelling and control. Firstly, this chapter introduces the basic operation of a typical modern wind turbine in Sects. 2.2 and 2.3 and followed by the modelling aspect of a wind turbine including aerodynamics, rotor, blades and tower in Sect. 2.4. Section 2.5 presents background of the model predictive control. Subsequently, the fatigue load assessment methods are discussed in Sect. 2.6. Section 2.7 presents the details of the simulation package FAST (Jonkman & Buhl Jr., 2005) and simulation turbine NREL 5 MW (Jonkman, Butterfield, Musial, & Scott, 2009). Finally, the design of the baseline CPC and IPC controller is discussed in Sect. 2.8 and followed by a summary in Sect. 2.9.

2.2 Modes of Operations

The power available from the wind is a function of the cube of the wind speed as follows (Pao & Johnson, 2009):

$$P_{\text{wind}}(t) := \frac{1}{2}\rho A v^3(t), \tag{2.1}$$

where P_{wind} denotes the power available from the wind, $\rho, A \in \mathbb{R}$ represent the air density and swept area of the rotor and v is the wind speed. Notice that in the introductory Sects. 2.2 and 2.3, the wind speed v is assumed to be the averaged wind speed across the rotor. The power available from the wind P_{wind} cannot be fully captured by a wind turbine, and there is a theoretical maximum aerodynamic efficiency of 0.593 on the available wind power that can be extracted by a wind turbine, where the maximum aerodynamic efficiency is known as the Betz limit

W. H. (Alan) Lio, *Blade-Pitch Control for Wind Turbine Load Reductions*, Springer Theses, https://doi.org/10.1007/978-3-319-75532-8_2

(Betz, 1966). Therefore, the power P captured by a wind turbine is typically defined as follows:

$$P(t) := \frac{1}{2}\rho A C_{\mathrm{p}}(\theta, \lambda) v^3(t). \tag{2.2}$$

This is the same as 2.1 except in 2.2, there is an additional term C_p, that is the aerodynamic efficiency which is a function of the pitch angle of the blades θ and tip speed ratio λ. Notice that in the introductory Sects. 2.2 and 2.3, the blade pitch angle θ is assumed to be adjusted collectively or the averaged pitch angle of three blades. The tip speed ratio is the ratio between the tangential speed of the blade tip and the actual speed of the wind, defined as follows:

$$\lambda(t) := \frac{\omega(t) r}{v(t)}, \tag{2.3}$$

where v and ω denote the wind speed and the rotor speed, respectively, whilst $r \in \mathbb{R}$ represents the total blade length.

A typical wind turbine operates in different modes of operation based on the available wind speed v. Figure 2.1 depicts the power curves for the power available from the wind P_{wind} and power captured by a typical modern wind turbine P. The modes of operation can be defined in terms of the wind speeds and these wind speeds are the cut-in v_{in}, cut-out v_{out} and rated wind speed v_{rated} respectively. For the NREL 5 MW simulation turbine (Jonkman et al., 2009), that employed in this thesis, the values for these wind speeds are as follows: $v_{\mathrm{in}} = 3\,\mathrm{ms}^{-1}$, $v_{\mathrm{out}} = 25\,\mathrm{ms}^{-1}$ and $v_{\mathrm{rated}} = 11.4\,\mathrm{ms}^{-1}$. As shown in Fig. 2.1, when the available wind speed is below the cut-in wind speed v_{in}, the wind turbine does not generate any power because it is not economical or environmentally friendly to do so, for example, studies by Arnett, Huso, Schirmacher, and Hayes (2011) suggested that increasing the cut-in wind speed can reduce the fatalities of wildlife such as birds and bats. When the wind condition is above the cut-out wind speed v_{out}, the wind turbine also stops producing power in order to prevent any damage to mechanical components and the generator. In

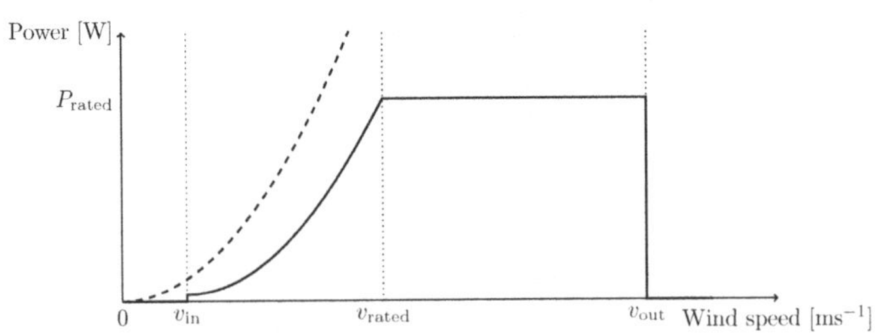

Fig. 2.1 Relationship of the wind power and wind speed. The wind power P_{wind} (dashed line) cannot be fully captured by the wind turbine. A power curve of a typical wind turbine P is denoted as solid line

below-rated wind conditions where the available wind is between the cut-in v_{in} and rated speed v_{rated}, the turbine operates at variable rotor speed in order to capture the maximum power from the wind. In contrast, in above-rated wind conditions where the wind is between the rated v_{rated} and cut-out speed v_{out}, the turbine limits its power generation at the rated power P_{rated} to prevent overloading the generator and other key mechanical components such as blades and main shaft bearing.

2.3 Control Objective

In a modern wind turbine, typical control systems are the yaw control, generator torque control and blade pitch control. The purpose of yaw controllers is to ensure the rotor is facing in the direction of the upcoming wind by rotating the turbine nacelle. Generator torque and blade pitch control are employed to regulate the rotor or generator speed. To understand how the generator torque and the blade pitch controller works, the power coefficient from (2.2) needs to be examined in detail.

Figure 2.2 shows the function of the power coefficient C_p against the tip speed ratio λ and the blade pitch angle θ. This function varies based on the turbine blade configurations and the function in Fig. 2.2 is obtained from the high-fidelity NREL 5 MW simulation turbine. Bearing in mind that the power of the wind turbine (2.2) is dependent on C_p. In below-rated wind conditions, wind turbines need to extract as much power as possible, thus, by inspection of Fig. 2.2, it is not difficult to understand that the optimal strategy is to keep the blade pitch at 0 degrees and the tip speed ratio around 7.8. To illustrate the ideas clearly, Fig. 2.3 shows the power coefficient curve when $\theta = 0$. Thus, in below-rated wind conditions, the typical generator torque controllers aim to track the optimal tip speed ratio by changing the reaction torque

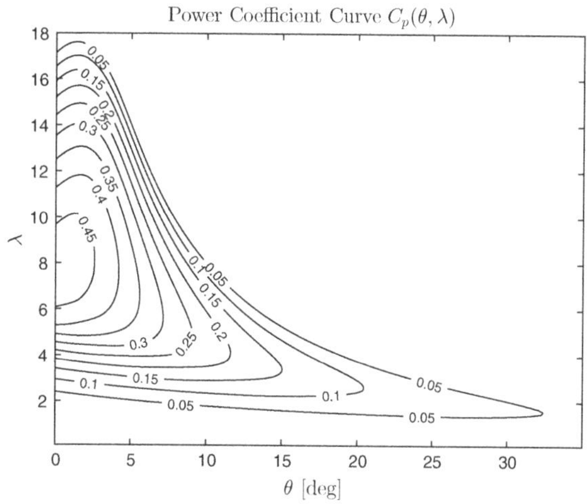

Fig. 2.2 Power coefficient C_p curve of the NREL 5 MW baseline wind turbine

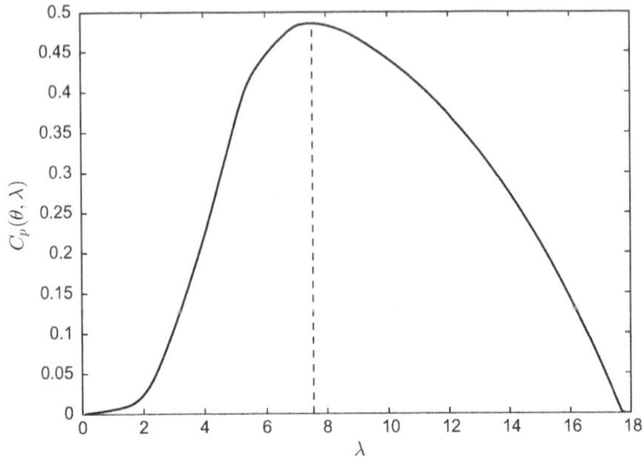

Fig. 2.3 Relationship of the power coefficient C_p against the tip speed ratio λ where the blade pitch angle θ are kept at 0°

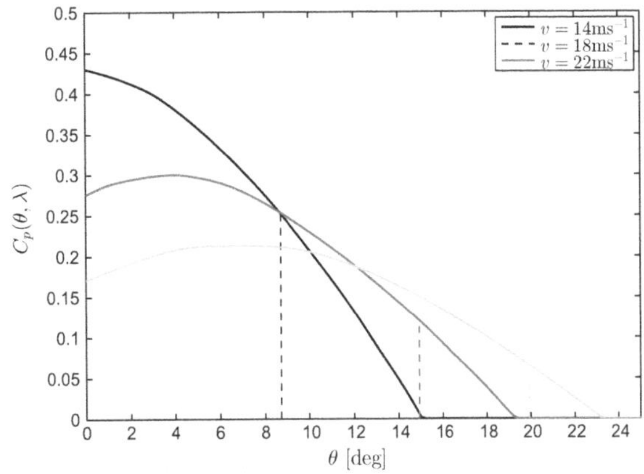

Fig. 2.4 Relationship of the power coefficient C_p against the blade pitch angle θ under various operating wind speed v

from the generator, whilst maintaining the blade pitch at constant (e.g. Laks, Pao, & Wright, 2009; Pao & Johnson, 2009).

In above-rated wind conditions, the power output reaches its rated value and the turbine needs to regulate its output. This is achieved by reductions in the power coefficient by changing the blade pitch angles. Typically, the reaction torque of the generator is fixed at its rated value and the blade pitch controller adjusts the pitch angles to alleviate the excessive aerodynamic loadings on the blades in order to regulate the rotor speed (Laks et al., 2009; Pao & Johnson, 2009). For any specific operating wind speeds v in the above-rated wind conditions, there is a corresponding fixed tip-speed ratio λ since the rotor speed ω is fixed at its rated value. This can be computed from Eq. (2.3). Thus, based on the different operating point λ corresponding to different operating wind speed v, the relationship between the C_p and blade pitch angle θ under different wind speed can be examined in Fig. 2.2. To illustrate clearly, Fig. 2.4 depicts how changes in blade pitch angle affect the power coefficient C_p under various operating wind speeds. In addition, for any specific operating wind conditions, there is a corresponding steady-state blade pitch angles such that the rotor speed is kept at rated. The dashed lines illustrate the steady states of the blade pitch angle.

Moreover, in the above-rated wind conditions, since the rotor is kept at its rated value $\omega_0 \in \mathbb{R}$, thus, based on Eq. (2.2), the power output can also be expressed in terms of aerodynamic torque τ_a that causes the rotor to spin as follows:

$$P(t) := \tau_a \omega_0 = \frac{1}{2}\rho A C_p(\theta, \lambda) v^3(t), \tag{2.4}$$

and the aerodynamic torque can be simplified as follows:

$$\tau_a(t) := \frac{1}{2}\rho A C_p(\theta, \lambda) v^3(t) \omega_0^{-1}. \tag{2.5}$$

Notice that under different operating wind speed, the gradients around the operating points vary slightly as shown in Fig. 2.4, which suggests that the static gain $\frac{d\tau_a}{d\theta} \in \mathbb{R}$ at each wind operating conditions changes slightly. More detail is discussed in Sect. 2.4.

Similar to the aerodynamic torque (2.5), the thrust on the rotor F_t is also a function of the tip speed ratio and the blade pitch angle, defined as follows:

$$F_t(t) := \frac{1}{2} A C_t(\theta, \lambda) v^2(t), \tag{2.6}$$

where C_t is the thrust coefficient. Figure 2.5 depicts the relationship between the turbine thrust, tip speed ratio λ and blade pitch angle θ. In above-rated wind conditions, the rotor is assumed to be at its rated value, thus, similar analysis can also be applied to the thrust. Figure 2.6 illustrates how changes in the blade pitch affect the rotor

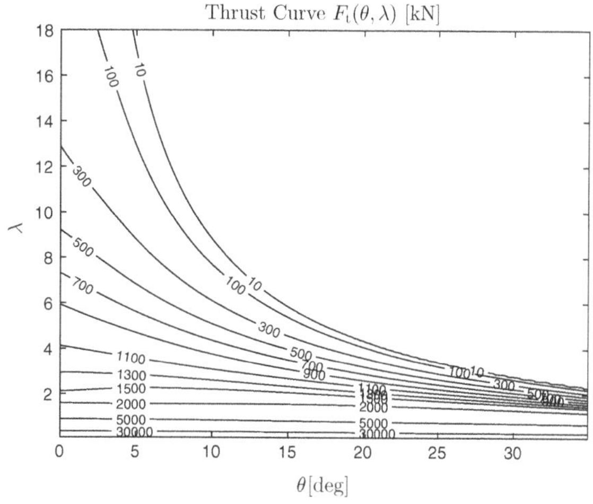

Fig. 2.5 Aerodynamic thrust curve of the NREL 5 MW baseline wind turbine

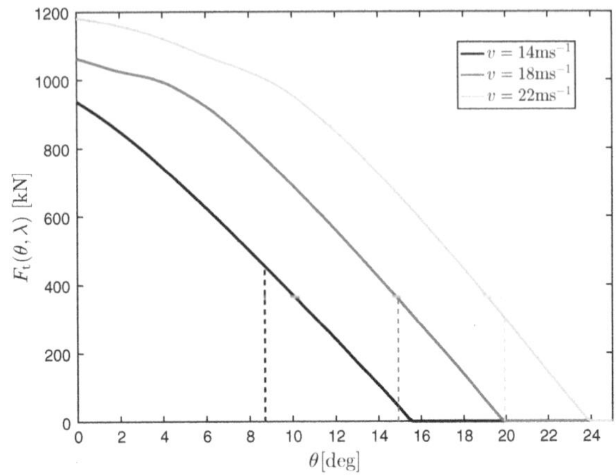

Fig. 2.6 Relationship of the rotor thrust F_t against the blade pitch angles θ under various operating wind speed v

thrust. Notice that the gradients $\frac{dF_t}{d\theta} \in \mathbb{R}$ at each operating points are quite similar. These gradients will be used in modelling in Sect. 2.4.

In most of the literature, the generator torque and blade pitch control loops are designed separately (e.g. Bossanyi, 2000; Laks et al., 2009), which is also favoured by the industry for the simplicity of implementations, and a switching logic is employed to handle the transitions between below and above-rated wind conditions. Nonetheless, there are some studies that used model predictive control to embed the switching logic as a part of the controller design (e.g. Henriksen, Hansen, & Poulsen, 2012; Soliman, Malik, & Westwick, 2010).

So far, the discussions in this section mainly focus on the generator torque control and blade collective pitch control. As mentioned in Chap. 1, blade individual pitch control (IPC) begins to be employed by an growing amount of studies, as the IPC offers a means for load attenuations on the turbine blade and rotor. Typically, in above-rated wind conditions, the pitch angle of each blade is adjusted individually in response to the measurements of flap-wise blade root bending moments. Notice that in this section, the pitch angle θ is defined as the collective pitch input, and the wind speed v is assumed to be the averaged wind speed across the rotor. In the following section, the pitch angle θ and wind speed v are defined in more detail.

2.4 Modelling of Wind Turbines

This section discusses some modelling aspects of wind turbines that are typically used in the blade-pitch control. Firstly, the aerodynamic loadings that exert torque and thrust on the rotor blades are discussed. It is followed by the dynamics of the rotor, blades and tower. Lastly, the models of actuator and filter dynamics are presented.

2.4.1 Aerodynamic Loadings

As mentioned in Sect. 1.2 in Chap. 1, the aerodynamic forces on the blades are generated by the air flow passing the aerofoil-shaped body of the blade. As shown in Fig. 2.7, typically, the aerofoil shape and angles of attack are different along the blade span-wisely, namely, from the blade root to blade tip. The aerofoil characteristics of the blade are optimised based on the resultant velocity that takes into account the tangential speed of the blade section and the oncoming wind speed (Burton, Jenkins, Sharpe, & Bossanyi, 2011). As a result, the aerofoil characteristics, namely the lift and drag coefficients, vary along the blade span-wise location. Therefore, to model the aerodynamic forces, blade element momentum (BEM) theory is widely used in wind turbine control (e.g. Burton et al., 2011; Geyler & Caselitz, 2008; Mirzaei, Henriksen, Poulsen, Niemann, & Hansen, 2012; Schlipf, Sandner, Raach, Matha, & Cheng, 2013), where the entire blade is divided into small blade elements span-wisely with the blade span-wise location index denoted as $l \in \{1, \ldots, L\} \subset \mathbb{Z}$. Subsequently, the in-plane $F_{i,l}^{\text{in}}$ and out-of-pane forces $F_{i,l}^{\text{out}}$ for each blade $i \in \{1, 2, 3\}$ at location l are defined as follows:

$$F_{i,l}^{\text{in}}(t) := \frac{1}{2}\rho A_l C_{\text{in}_l}(\theta_i, \lambda) v_{i,l}^2(t), \tag{2.7a}$$

$$F_{i,l}^{\text{out}}(t) := \frac{1}{2}\rho A_l C_{\text{out}_l}(\theta_i, \lambda) v_{i,l}^2(t). \tag{2.7b}$$

where A_l denotes the area of the blade element l and $v_{i,l}$ represents the wind speed experienced by the blade element l of blade i. The coefficients of in-plane and out-of-plane forces of blade element l, C_{in_l} and C_{out_l}, are functions of the pitch angle θ_i of blade i and tip-speed ratio λ, which is defined as $\lambda := \frac{\omega r}{v}$ where v is the operating wind speed that is the averaged wind speed across the rotor and the rotor speed ω, which is at the rated speed $\omega_0 \in \mathbb{R}$ in above-rated wind conditions.

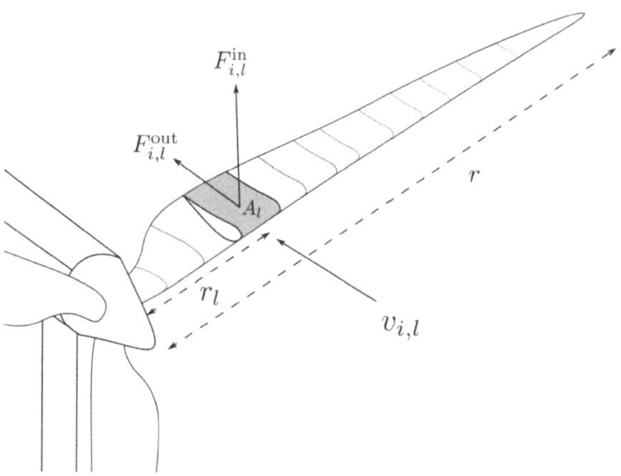

Fig. 2.7 A turbine blade is assumed to be divided into smaller blade elements. A pair of in-plane and out-of-plane forces, $F_{i,l}^{\text{in}}$ and $F_{i,l}^{\text{out}}$, is generated by the blade element l of blade i with shaded area of A_l passing through the airflow with speed of $v_{i,l}$. The blade length between three blade root to blade element is denoted as r_l and r is the total blade length

Notice that the in-plane and out-of-plane forces are non-linear functions (2.7). In this thesis, a linear model is employed, where these linear system models typically exhibit properties that are much simpler than the non-linear form. The main reason is that the non-linear function of in-plane and out-of-plane forces (2.7) can be approximated as a linear function at an operating point, because, in the above-rated wind conditions, the rotor operates at the rated speed. In addition, there exists a substantial amount of literature for linear system design and analysis. Thus, this justifies the use of linear models. The in-plane and out-of-plane forces can be linearised around the operating conditions by the use of Taylor series expansion as follows:

$$F_{i,l}^{\text{in}}(t) \approx \left.\frac{dF_{i,l}^{\text{in}}}{d\theta}\right|_{\theta^*,\omega_0,v} \theta_i(t) + \left.\frac{dF_{i,l}^{\text{in}}}{dv}\right|_{\theta^*,\omega_0,v} v_{i,l}(t), \tag{2.8a}$$

$$F_{i,l}^{\text{out}}(t) \approx \left.\frac{\partial F_{i,l}^{\text{out}}}{d\theta}\right|_{\theta^*,\omega_0,v} \theta_i(t) + \left.\frac{dF_{i,l}^{\text{out}}}{dv}\right|_{\theta^*,\omega_0,v} v_{i,l}(t). \tag{2.8b}$$

The linearised forces are computed under an assumption that the wind across the rotor is spatially uniformly distributed. At each operating wind speed v, there is a corresponding steady-state of the blade pitch angle θ^* that regulates the rotor at the rated speed ω_0. Small deviations of blade pitch and wind speed around their steady-states are denoted as θ_i and $v_{i,l}$, respectively. Notice that the variations of in-plane and out-plane with respect to blade pitch angle, $\frac{dF_{i,l}^{\text{in}}}{d\theta}, \frac{dF_{i,l}^{\text{out}}}{d\theta} \in \mathbb{R}$, and with respect to wind speed, $\frac{dF_{i,l}^{\text{in}}}{dv}, \frac{dF_{i,l}^{\text{out}}}{dv} \in \mathbb{R}$, along the blade element l can be obtained from the simulation turbine as shown in Figs. 2.8 and 2.9. These results were generated from the response of the blade forces to a step change in the blade pitch angles or wind speed under a set of operating conditions. Notice that the blade is divided into blade segments of $L = 17$ in this thesis which is adopted from the simulation turbine (Jonkman & Buhl Jr., 2005).

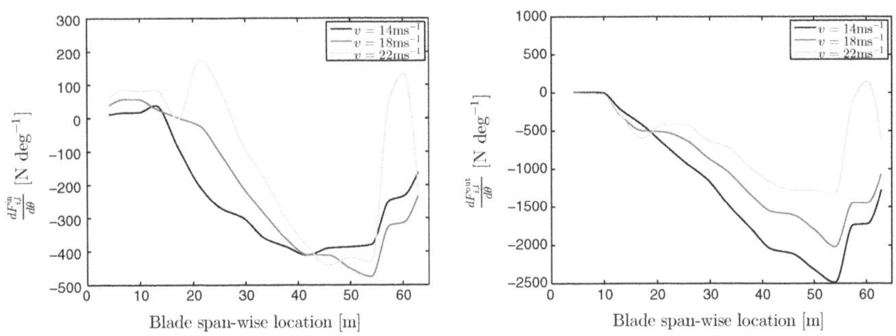

(A) Variations of in-plane force with respect to blade pitch angle.

(B) Variations of out-of-plane force with respect to blade pitch angle.

Fig. 2.8 Variations of in-plane and out-of-plane forces to blade pitch angle along the blade span-wise location under different operating wind conditions

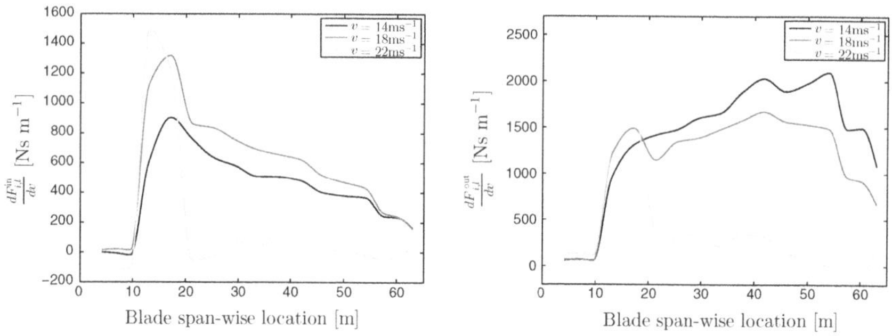

(A) Variations of in-plane force with respect to wind speed.

(B) Variations of out-of-plane force with respect to wind speed.

Fig. 2.9 Variations of in-plane and out-of-plane forces to wind speed along the blade span-wise location under different operating wind conditions

With these simplified in-plane and out-of-plane blade forces, the aerodynamic torque τ_a and thrust F_t on the rotor and flap-wise aerodynamic loadings on the blade M_{flap_i} [1] can be modelled as follows:

$$\tau_\text{a}(t) := \sum_{i=1}^{3}\sum_{l=1}^{L} F_{i,l}^{\text{in}}(t) r_l, \tag{2.9a}$$

$$F_\text{t}(t) := \sum_{i=1}^{3}\sum_{l=1}^{L} F_{i,l}^{\text{out}}(t), \tag{2.9b}$$

$$M_{\text{flap}_i}(t) := \sum_{l=1}^{L} F_{i,l}^{\text{out}}(t) r_l, \tag{2.9c}$$

where r_l denotes the blade length from blade root to blade span-wise location l (See Fig. 2.7). Consequently, substituting (2.8) into (2.9) yields:

$$\tau_\text{a}(t) = \left.\frac{d\tau_\text{a}}{d\theta}\right|_{\theta^*,\omega_0,v} \bar{\theta}(t) + \tau_\text{a}^\text{d}(t), \tag{2.10a}$$

$$F_\text{t}(t) = \left.\frac{dF_\text{t}}{d\theta}\right|_{\theta^*,\omega_0,v} \bar{\theta}(t) + F_\text{t}^\text{d}(t), \tag{2.10b}$$

$$M_{\text{flap}_i}(t) = \left.\frac{dM_\text{flap}}{d\theta}\right|_{\theta^*,\omega_0,v} \theta_i(t) + M_i^\text{d}(t), \tag{2.10c}$$

where $\bar{\theta} := \frac{1}{3}\sum_{i=1}^{3}\theta_i$ is the collective pitch angle that is the average of the pitch angles θ_i of three blades and the rest of the variables are defined as follows:

[1] The flap-wise aerodynamic loadings is a sum of the out-of-plane forces multiplied by the corresponding radial distance from the blade root.

$$\left.\frac{d\tau_{\mathrm{a}}}{d\theta}\right|_{\theta^*,\omega_0,v} := \sum_{i=1}^{3}\sum_{l=1}^{L}\left.\frac{dF_{i,l}^{\mathrm{in}}}{d\theta}\right|_{\theta^*,\omega_0,v} r_l, \quad \tau_{\mathrm{a}}^{\mathrm{d}}(t) := \sum_{i=1}^{3}\sum_{l=1}^{L}\left.\frac{dF_{i,l}^{\mathrm{in}}}{dv}\right|_{\theta^*,\omega_0,v} v_{i,l}(t) r_l, \tag{2.10d}$$

$$\left.\frac{dF_{\mathrm{t}}}{d\theta}\right|_{\theta^*,\omega_0,v} := \sum_{i=1}^{3}\sum_{l=1}^{L}\left.\frac{dF_{i,l}^{\mathrm{out}}}{d\theta}\right|_{\theta^*,\omega_0,v}, \quad F_{\mathrm{t}}^{\mathrm{d}}(t) := \sum_{i=1}^{3}\sum_{l=1}^{L}\left.\frac{dF_{i,l}^{\mathrm{out}}}{dv}\right|_{\theta^*,\omega_0,v} v_{i,l}(t), \tag{2.10e}$$

$$\left.\frac{dM_{\mathrm{flap}}}{d\theta}\right|_{\theta^*,\omega_0,v} := \sum_{l=1}^{L}\left.\frac{dF_{i,l}^{\mathrm{out}}}{dl\theta}\right|_{\theta^*,\omega_0,v} r_l, \quad M_i^{\mathrm{d}}(t) := \sum_{l=1}^{L}\left.\frac{dF_{i,l}^{\mathrm{out}}}{dv}\right|_{\theta^*,\omega_0,v} v_{i,l}(t) r_l, \tag{2.10f}$$

where $\tau_{\mathrm{a}}^{\mathrm{d}}$, $F_{\mathrm{t}}^{\mathrm{d}}$, $M_i^{\mathrm{d}} \in \mathbb{R}$ is denoted as the wind-induced torque disturbance, thrust disturbance and blade moment disturbance, respectively. Despite the variations in model parameters, these variations are not significant, hence, adaptive control or gain-scheduling design are not considered in this thesis. Instead, this thesis ensures the feedback control loop is robust to these parameter uncertainties by checking their robust stability margin. In this thesis, the linear model is obtained from linearising the aerodynamic forces (2.7) at the operating condition $v = 18\,\mathrm{ms}^{-1}$, chosen since this value is near the centre of the range of wind speeds covering above-rated wind conditions. Thus, for brevity, the variations in (2.10), for example, $\left.\frac{d\tau_{\mathrm{a}}}{d\theta}\right|_{\theta^*,\omega_0,v}$, will be expressed in a simpler form, for example, $\frac{d\tau_{\mathrm{a}}}{d\theta}$, in the rest of this thesis. One exception is Chap. 7, where in order to ensure better estimations of wind-induced blade disturbance for feed-forward control, the disturbance is calculated based on the various wind conditions.

Remark 2.1 An interesting observation is that the static gain $\frac{dF_{\mathrm{t}}}{d\theta} = 46\,\mathrm{kN\,deg}^{-1}$ at $v = 18\,\mathrm{ms}^{-1}$, which can be calculated based on (2.10e) with data from Fig. 2.8b. From the aerodynamic thrust curve F_{t} in Fig. 2.6, the gradient also reveals that $\frac{dF_{\mathrm{t}}}{d\theta} = 46\,\mathrm{kN\,deg}^{-1}$. These consistent values of the static gains, that one obtained from simulations and anther one from the manual (Jonkman & Buhl Jr., 2005), validates the modelling approach for the aerodynamic loadings, discussed in this section.

Remark 2.2 The flap-wise blade root bending moment is a term that is widely used to describe the blade bending moment in the direction perpendicular to the rotor plane, for example, Wang, Johnson, Wright, and Wright (2012), Castaignet, Barlas, and Buhl (2013) and Lu, Bowyer, and Jones (2015). It also sometimes is termed as *the out-of-plane blade root bending moment* in other works, for example, Bossanyi (2003a) and Dunne, Pao, Wright, Jonkman, and Kelley (2011).

2.4.2 *Structural Dynamics*

Three structural dynamics are considered in this thesis: dynamics of the rotor, tower and blade. Notice that the structural dynamics are approximated as second-order, where the dynamics at higher frequencies are neglected. The main reason is that the controller synthesised from this low-order linear model can be made to be insensitive to the neglected dynamics, justifying the use of the second-order model for feedback controller design.

2.4.2.1 Rotor Dynamics

The rotational speed of the rotor is mainly driven by the in-plane forces or aerodynamic torque on the rotor. Typically, the dynamics of the rotor speed ω is modelled as follows:

$$J_{\mathrm{r}}\dot{\omega}(t) + \zeta_{\mathrm{r}}\omega(t) = \tau_{\mathrm{a}}(t) - \tau_{\mathrm{e}}(t), \tag{2.11a}$$

where $J_{\mathrm{r}}, \zeta_{\mathrm{r}} \in \mathbb{R}$ denote the moment of inertia of the rotor and aerodynamic damping coefficient, respectively, and $\tau_{\mathrm{e}} \in \mathbb{R}$ is the reaction torque from the generator, which is kept constant in the above-rated wind conditions. Substituting the aerodynamic torque from (2.10), the model is simplified as follows:

$$J_{\mathrm{r}}\dot{\omega}(t) + \zeta_{\mathrm{r}}\omega(t) = \frac{d\tau_{\mathrm{a}}}{d\theta}\bar{\theta}(t) + \tau_{\mathrm{a}}^{\mathrm{d}}(t). \tag{2.11b}$$

Notice that it is assumed that the drive train is rigid, hence, the dynamics of the drive train is not considered in this thesis.

2.4.2.2 Blade Dynamics

The fatigue loadings on the blades are typically characterised by the blade flap-wise root bending moments (Bossanyi, 2000). The dynamical response of the flap-wise blade root bending moment M_i for blade i to the flap-wise aerodynamic loadings M_{flap_i} can be modelled as a second-order system:

$$\ddot{M}_i(t) + 2\pi f_{\mathrm{b}}\zeta_{\mathrm{b}}\dot{M}_i(t) + (2\pi f_{\mathrm{b}})^2 M_i(t) = (2\pi f_{\mathrm{b}})^2 M_{\mathrm{flap}_i}(t), \tag{2.12a}$$

where $f_{\mathrm{b}}, \zeta_{\mathrm{b}} \in \mathbb{R}$ denote the natural frequency of the blade's first flap-wise mode and damping ratio, respectively. Substituting the flap-wise aerodynamic loadings from (2.10), the blade model (2.12a) becomes:

$$\ddot{M}_i(t) + 2\pi f_b \zeta_b \dot{M}_i(t) + (2\pi f_b)^2 M_i(t) = (2\pi f_b)^2 \Big(\frac{dM_{\text{flap}}}{d\theta} \theta_i(t) + M_i^{\text{d}}(t) \Big). \tag{2.12b}$$

Notice that the gravitational effect and centrifugal stiffening of the rotor blades are not considered in the blade model (2.12).

2.4.2.3 Tower Dynamics

The fore-aft motion of the tower is mainly driven by the aerodynamic thrust on the rotor. Based on first principles, the tower dynamics is modelled as a mass-spring-damper system, that is defined as follows (e.g. Selvam, Kanev, van Wingerden, van Engelen, & Verhaegen, 2009):

$$m_t \ddot{x}_{\text{fa}}(t) + d_t \dot{x}_{\text{fa}}(t) + k_t x_{\text{fa}}(t) = F_t(t) - \frac{3}{2h} M_{\text{tilt}}(t), \tag{2.13a}$$

where x_{fa} denotes the fore-aft displacement of the tower-top, whilst $m_t, d_t, k_t \in \mathbb{R}$ denote the equivalent tower mass, damping and stiffness coefficient, respectively, and M_{tilt} is the tilt moment of the rotor, that is defined in Sect. 3.2 in Chap. 3. Based on an assumption that the fore-aft tower motion is approximated by the motion of a prismatic beam, the notion between tilting rotation and fore-aft displacement of the tower-top is $\frac{2}{3h}$, where $h \in \mathbb{R}$ is the height of the tower (Selvam et al., 2009; van Engelen & van der Hooft, 2005). More detail is discussed in Chap. 5. By expressing the aerodynamic thrust in a linear form (2.10), the tower model (2.13a) becomes:

$$m_t \ddot{x}_{\text{fa}}(t) + d_t \dot{x}_{\text{fa}}(t) + k_t x_{\text{fa}}(t) = \frac{dF_t}{d\theta} \bar{\theta}(t) + F_t^{\text{d}}(t) - \frac{3}{2h} M_{\text{tilt}}(t). \tag{2.13b}$$

2.4.2.4 Couplings Between Tower, Blade and Rotor

The effect of the tower dynamics are not considered in the rotor (2.11) and blade models (2.12) because the feedback blade pitch controllers are deliberately tuned to be insensitive to the couplings from the tower by shaping the open-loop frequency response to have a low gain at the natural frequency of tower's first fore-aft mode. Details are discussed in Chap. 4.

However, there is one exception in Chap. 5 since the couplings need to be considered for tower estimation and control problems. The apparent stream-wise wind speeds experienced by the turbine blades and rotor are influenced by the motion of the tower. Typically, the apparent stream-wise wind speed $v_{i,l}$ experienced by blade element l of blade i takes into account the actual stream-wise wind speed $v_{0_{i,l}}$ and the upwind motion of the blade caused by tower bending, that is described by the fore-aft velocity $\dot{x}_{\text{fa}}$ and fore-aft rotational velocity $\dot{\varphi}_{\text{fa}}$ of the tower-top. The stream-wise wind speed $v_{i,l}$ is defined as follows (e.g. Selvam et al., 2009; van Engelen & van der Hooft, 2005):

$$v_{i,l}(t) := v_{0_{i,l}}(t) - \dot{x}_{\mathrm{fa}}(t) + r_l \dot{\varphi}_{\mathrm{fa}}(t) \sin \phi_i(t), \tag{2.14a}$$

where ϕ_i is the azimuth angle of blade i, that is defined as the angle of the first blade from the horizontal yaw axis, rotating clock-wisely. As discussed above, the fore-aft rotation of the tower-top can be expressed as displacement as $\dot{\varphi}(t) := \frac{3}{2h}\dot{x}_{\mathrm{fa}}(t)$. Hence, the apparent stream-wise wind speed can be simplified as follows:

$$v_{i,l}(t) := v_{0_{i,l}}(t) + \left(\frac{3}{2h} r_l \sin \phi_i(t) - 1\right)\dot{x}_{\mathrm{fa}}(t). \tag{2.14b}$$

More detail is provided in Chap. 5.

2.4.3 *Actuator and Filter Dynamics*

2.4.3.1 Actuator Dynamics

The types of blade pitch actuators used in wind turbines can be divided into hydraulic or electrical servo systems. Hydraulic driving systems are commonly employed in wind turbines (Chiang, 2011). This is because unlike electrical servos, motor and gears are not used in hydraulic systems, thus, such properties obviate the problems of erosion and wear of gears over time. Studies of blade pitch actuators can be found in (Chiang, 2011) and the references therein. In this thesis, the dynamics of the actuator is modelled as a first-order model:

$$\tau_{\mathrm{act}} \dot{\theta}_i(t) + \theta_i(t) = \theta_i^{\mathrm{c}}(t), \tag{2.15}$$

where $\tau_{\mathrm{act}} \in \mathbb{R}$ denotes the time constant of the pitch actuator, whilst θ_i^{c} is the blade pitch command signal.

2.4.3.2 Filter Dynamics

In practice, the measurements of the rotor speed or flap-wise blade root moments are coupled with noise. Typically, filters are employed in wind turbines to remove any unwanted components before being used by the feedback controllers. For example, to decouple the rotor speed regulation loop from the tower control and IPC loop, a low-pass filter, with a cut-off frequency below the rotational frequency of the blade, is widely used for the measurement of the rotor speed. Similarly, a band-pass filter is employed on measurement signals from the blade load sensor in order to remove low-frequency drift and high-frequency noise. These low-pass $G_{\mathrm{lp}} \in \mathcal{R}$ and band-pass filters $G_{\mathrm{bp}} \in \mathcal{R}$ are defined as follows:

$$G_{\mathrm{lp}}(s) := \frac{2\pi f_{\mathrm{lp}}}{s + 2\pi f_{\mathrm{lp}}}, \tag{2.16a}$$

$$G_{\mathrm{bp}}(s) := \frac{2\pi f_{\mathrm{h}} s}{s^2 + 2\pi(f_{\mathrm{h}} + f_{\mathrm{l}})s + 4\pi^2 f_{\mathrm{h}} f_{\mathrm{l}}}, \tag{2.16b}$$

where $f_{\mathrm{lp}} \in \mathbb{R}$ denotes the cut-off frequency of the low-pass filter, whilst $f_{\mathrm{l}}, f_{\mathrm{h}} \in \mathbb{R}$ represent the upper and lower cut-off frequencies of the band-pass filter, respectively.

2.5 Model Predictive Control

As discussed in Chap. 1, one of the aims in this thesis is to investigate the use of advance knowledge of the upcoming wind for improving the load attenuation performance of the blade-pitch controller. Given that model predictive control (MPC) can systematically incorporate the advance information into the optimisation process, that also takes into account of the constraints of the pitch actuators, thus, it is favoured by many applications and also employed in this thesis.

The working principle of a standard MPC is described briefly as follows. The control algorithm uses a mathematical model together with past measurements of input and output (and future information if available) to construct predictions. Based on these predictions, a future control input sequence is computed that ensures all the system states satisfying the constraint requirements. Subsequently, only the first control input of the sequence is implemented to the plant. The optimisation process is repeated for the next sample. Since the repetition process of the MPC algorithm is a form of feedback, hence, this feedback property gives a certain degree of inherent robustness to the uncertainties that arise from the mismatch between predictions of the mathematical model and actual behaviours of the system plant.

2.5.1 Generalised Predictive Control

The generalised predictive control (GPC) is one of the most popular predictive control algorithms in academia (Valencia-Palomo, 2010) and it was originated by Clarke, Mohtadi, and Tuffs (1987a, 1987b). A typical linear discrete-time state-space form of a system model is defined as follows:

$$\begin{aligned} \zeta_{k+1} &= \mathcal{A}\zeta_k + \mathcal{B}u_k, \\ y_k &= \mathcal{C}\zeta_k, \end{aligned} \tag{2.17}$$

where $\zeta \in \mathbb{R}^{n_\zeta}$, $y \in \mathbb{R}^{n_y}$, $u \in \mathbb{R}^{n_u}$ denote the state, output and input of the model, whilst $\mathcal{A} \in \mathbb{R}^{n_\zeta \times n_\zeta}$, $\mathcal{B} \in \mathbb{R}^{n_\zeta \times n_u}$, $\mathcal{C} \in \mathbb{R}^{n_y \times n_\zeta}$ represent the state, input and output matrix, respectively. The system (2.17) is assumed to be controllable and observable.

It is convenient for formulating the offset-free cost function if the state-space model (2.17) is defined in terms of input increment $\Delta u_k \in \mathbb{R}^{n_u}$ as follows:

$$\underbrace{\begin{bmatrix} \zeta_{k+1} \\ u_k \end{bmatrix}}_{x_{k+1}} = \underbrace{\begin{bmatrix} \mathcal{A} & \mathcal{B} \\ 0 & I \end{bmatrix}}_{A} \underbrace{\begin{bmatrix} \zeta_k \\ u_{k-1} \end{bmatrix}}_{x_k} + \underbrace{\begin{bmatrix} \mathcal{B} \\ I \end{bmatrix}}_{B} \Delta u_k,$$
$$y_k = \underbrace{\begin{bmatrix} \mathcal{C} & 0 \end{bmatrix}}_{C} x_k. \tag{2.18}$$

The prediction sequence of the state $\underset{\to k}{x} \in \mathbb{R}^{n_x n}$ for prediction horizon $n \in \mathbb{R}$ can be represented as follows:

$$\underbrace{\begin{bmatrix} x_{1|k} \\ x_{2|k} \\ x_{3|k} \\ \vdots \\ x_{n|k} \end{bmatrix}}_{\underset{\to k}{x}} = \underbrace{\begin{bmatrix} A \\ A^2 \\ A^3 \\ \vdots \\ A^n \end{bmatrix}}_{P_{xx}} x_{0|k} + \underbrace{\begin{bmatrix} B & 0 & 0 & \cdots \\ AB & B & 0 & \cdots \\ A^2B & AB & B & \cdots \\ \vdots & \vdots & \vdots & \vdots \\ A^{n-1}B & A^{n-2}B & A^{n-3}B & \cdots \end{bmatrix}}_{H_x} \underbrace{\begin{bmatrix} \Delta u_{0|k} \\ \Delta u_{1|k} \\ \Delta u_{2|k} \\ \vdots \\ \Delta u_{n-1|k} \end{bmatrix}}_{\Delta \underset{\to k-1}{u}}, \tag{2.19a}$$

where $x_{i|k}$ denotes the prediction of x_{k+i} evaluated at time step k, whilst $\Delta \underset{\to k-1}{u}$ is the future input sequence. Notice that $x_{0|k}$ is x_k. Similarly, the prediction sequence of the output $\underset{\to k}{y} \in \mathbb{R}^{n_y n}$ is as follows:

$$\underbrace{\begin{bmatrix} y_{1|k} \\ y_{2|k} \\ y_{3|k} \\ \vdots \\ y_{n|k} \end{bmatrix}}_{\underset{\to k}{y}} = \underbrace{\begin{bmatrix} CA \\ CA^2 \\ CA^3 \\ \vdots \\ CA^n \end{bmatrix}}_{P} x_{0|k} + \underbrace{\begin{bmatrix} CB & 0 & 0 & \cdots \\ CAB & CB & 0 & \cdots \\ CA^2B & CAB & CB & \cdots \\ \vdots & \vdots & \vdots & \vdots \\ CA^{n-1}B & CA^{n-2}B & CA^{n-3}B & \cdots \end{bmatrix}}_{H} \Delta \underset{\to k-1}{u}. \tag{2.19b}$$

Typically, the cost function employed in MPC is described by a quadratic function that penalises the deviation of the control variables and the predicted error between the output and desired set-point, as follows:

$$\begin{aligned} J_k &= \sum_{i=1}^{n} \Big((r_{i|k} - y_{i|k})^T Q (r_{i|k} - y_{i|k}) + \Delta u_{i-1|k}^T R \Delta u_{i-1|k} \Big), \\ &= (\underset{\to k}{r} - \underset{\to k}{y})^T \bar{Q} (\underset{\to k}{r} - \underset{\to k}{y}) + \Delta \underset{\to k-1}{u}^T \bar{R} \Delta \underset{\to k-1}{u}, \end{aligned} \tag{2.20a}$$

where $r_{i|k}$ denotes the set-point, whilst $Q \in \mathbb{R}^{n_y \times n_y}$, $R \in \mathbb{R}^{n_u \times n_u}$ are the weighting matrices that specify the penalties on the error and input, respectively. Notice that (2.20a) is a offset-free cost function since in the steady-state, there is zero tracking error if the following conditions hold:

$$y_k = r_k, \;\; \Delta u_k = 0. \tag{2.20b}$$

By performing a minimisation of the cost function (2.20a), the unconstrained optimal control sequence can be computed:

$$\Delta \underset{\rightarrow k-1}{u} = \arg \min_{\Delta \underset{\rightarrow k-1}{u}} J_k. \tag{2.20c}$$

Given that the cost function is quadratic, there exists a unique minimum which can be found by setting the first derivative to zero as follows:

$$\frac{dJ_k}{d\Delta \underset{\rightarrow k-1}{u}} = 2(H^T\bar{Q}H + \bar{R})\Delta \underset{\rightarrow k-1}{u} + 2H^T\bar{Q}\underset{\rightarrow k}{r} - Px_{0|k}], \tag{2.20d}$$

$$\frac{dJ_k}{d\Delta \underset{\rightarrow k-1}{u}} = 0 \Rightarrow \Delta \underset{\rightarrow k-1}{u} = (H^T\bar{Q}H + \bar{R})^{-1}H^T\bar{Q}[\underset{\rightarrow k}{r} - Px_{0|k}]. \tag{2.20e}$$

The first input of the unconstrained optimal input sequence $\Delta \underset{\rightarrow k-1}{u}$ is implemented into the plant:

$$\Delta u_k = E\Delta \underset{\rightarrow k-1}{u} = \underbrace{E(H^T\bar{Q}H + \bar{R})^{-1}H^T\bar{Q}}_{P_r}\underset{\rightarrow k}{r} - \underbrace{E(H^T\bar{Q}H + R)^{-1}H^T\bar{Q}P}_{K}x_{0|k}, \tag{2.21}$$

where $E = \begin{bmatrix} I & 0 & \cdots & 0 \end{bmatrix} \in \mathbb{R}^{n_u \times n_u n}$ and $x_{0|k} = x_k$.

Notice that so far the input horizon n_c is assumed to be equal to prediction horizon n in the cost function (2.20a). If $n_c \neq n$, then the H matrix in the prediction (2.19) needs to be replaced with the following matrix:

$$H_c = \begin{bmatrix} B & 0 & \cdots & 0 \\ AB & B & \cdots & 0 \\ \vdots & \vdots & \vdots & \vdots \\ A^{n-1}B & A^{n-2}B & \cdots & A^{n-n_c}B \end{bmatrix}. \tag{2.22}$$

For constrained MPC algorithm, the process of minimisation of the cost function (2.20a) needs to take into account constraints, where typical constraints on the input rate, state and output are defined as follows:

$$\Delta u_{\min} \le \Delta u_{i|k} \le \Delta u_{\max}, \quad i = 1, \cdots, n, \tag{2.23a}$$
$$x_{\min} \le x_{i|k} \le x_{\max}, \quad i = 1, \cdots, n, \tag{2.23b}$$
$$y_{\min} \le y_{i|k} \le y_{\max}, \quad i = 1, \cdots, n, \tag{2.23c}$$

where $x_{\min}$, $\Delta u_{\min}$, $y_{\min}$ are the minimum values of the state, input rate and output, whilst $x_{\max}$, $\Delta u_{\max}$, $y_{\max}$ are the maximum values of the state, input and output, respectively. Notice that the constraint on the input is implicitly embedded to the state constraint since the input is augmented as a part of the state x_k in (2.18). By substituting (2.19) into (2.23), the constraints (2.23) can be combined into one single set of linear inequalities of the following form:

$$\underbrace{\begin{bmatrix} I \\ -I \\ H_x \\ -H_x \\ H \\ -H \end{bmatrix}}_{H_u} \Delta \underset{\rightarrow k-1}{u} \le \underbrace{\begin{bmatrix} u_{\max}\mathbf{1} \\ -u_{\min}\mathbf{1} \\ x_{\max}\mathbf{1} - P_{xx}x_{0|k} \\ -x_{\min}\mathbf{1} + P_{xx}x_{0|k} \\ y_{\max}\mathbf{1} - Px_{0|k} \\ -y_{\min}\mathbf{1} + Px_{0|k} \end{bmatrix}}_{h_u}, \tag{2.23d}$$

where $\mathbf{1} := [1, 1, \cdots, 1]^T$.

Consequently, the GPC algorithm can be summarised as follows:

Algorithm 2.5.1 (*GPC*) At every sample,

1. Perform the following optimisation:

$$\begin{aligned} &\arg \min_{\Delta \underset{\rightarrow k-1}{u}} \; \Delta \underset{\rightarrow k-1}{u}^T (H^T Q H + R) \Delta \underset{\rightarrow k-1}{u} + \Delta \underset{\rightarrow k-1}{u}^T H^T [\underset{\rightarrow k}{r} - Px_k], \\ &\text{s.t.} \quad H_u \Delta \underset{\rightarrow k-1}{u} \le h_u \end{aligned} \tag{2.24}$$

2. The first sample of the future input sequence $\Delta \underset{\rightarrow k-1}{u}$ is applied into the plant .

Notice that the optimisation problem (2.24) in Algorithm 2.5.1 is in a general quadratic program (QP) form, which is easy to solve (e.g. using `quadprog.m` in Matlab®).

Despite the GPC and its variants have been successfully and widely adopted in many industrial applications since 1970s, the issues of tunings and stability were major problems with GPC. For example, tuning of n_u and n_y is not an easy task and an excellent source that explains the effect of different horizon, n_u and n_y, can be found in Rossiter (2003). Although given such tuning guidelines, only a posteriori stability checks can be performed for GPC, where the nominal (unconstrained) control law is computed, then the corresponding closed-loop poles were examined. In the early 1990s, the stability issue of GPC was addressed by the concept of *dual-mode MPC*, that makes use of the celebrated linear quadratic optimal control results of the 1960s (Clarke & Scattolini, 1991; Kouvaritakis, Rossiter, & Schuurmans, 2000; Mosca & Zhang, 1992; Rawlings & Muske, 1993).

2.5.2 *Dual-Mode MPC*

Compared to GPC, dual-mode MPC offers guaranteed nominal stability a priori. Before introducing the dual-mode MPC, the background of the linear quadratic regulator (LQR) problem is provided.

2.5.2.1 Linear Quadratic Regulator

Consider the classical LQR problem with a quadratic cost function of the form:

$$\min_{\Delta u_k} J := \sum_{k=0}^{\infty} \left(x_k^T Q_x x_k + \Delta u_k^T R \Delta u_k \right). \tag{2.25}$$

The solution of (2.25) can be obtained by using Dynamic Programming and the recursive Bellman equation (Bellman, 1957), defined as follows:

$$\Delta u_k = -K x_k, \tag{2.26}$$

where K is a stabilising feedback controller such that the eigenvalues of $A - BK$ are inside the unit circle if the system model (2.17) is controllable and observable, and the matrix $Q_x \geq 0$ is real and non-negative definite and $R > 0$ is positive definite (Bellman, 1957). Notice that the GPC cost function (2.20a) with infinite prediction horizon ($n = \infty$) is equivalent to the LQR cost function (2.25) if $r_{i|k}$ is assumed to be zero and $Q_x = C^T Q C$. Then, the unconstrained control law of the GPC (2.21) simply becomes the LQR controller (2.26). However, when constraints (2.23) are presented, the minimisation of (2.25) is required at every sample with the plant current state $x_{0|k}$, then the problem becomes infinite-dimensional, which is intractable. Thus, the use of dual-mode paradigm enables the constrained optimisation of (2.25) to be specified as a finite-dimensional problem.

Notice that the cost function (2.25) assumed the set-point is zero $r_{i|k} = 0$. If $r_{i|k} \neq 0$, in order to ensure offset-free tracking, a steady-state target calculator is needed (Pannocchia & Rawlings, 2003). Consider the control law (2.26), that drives the state x_k to the steady-state x_k^s, is defined as follows:

$$\Delta u_k = -K(x_k - x_k^s) + \Delta u_k^s, \tag{2.27}$$

where Δu_k^s is the steady-state input. To ensure offset-free tracking, a condition is needed to be held, where if $x_k = x_k^s$ and $\Delta u_k = \Delta u_k^s$, then $y_k^s = C x_k^s = r_k$. Thus, in steady-state, the state-space model can be defined as follows:

$$\begin{aligned} x_{k+1}^s &= Ax_k^s + B\Delta u_k^s, \\ y_k^s &= Cx_k^s = r_k, \end{aligned} \tag{2.28}$$

where the steady-state input Δu_k^s and state x_k^s can be computed by the steady-state target calculator (Pannocchia & Rawlings, 2003), that is defined as follows:

$$\begin{bmatrix} x_k^s \\ \Delta u_k^s \end{bmatrix} := \begin{bmatrix} C & 0 \\ I - A & -B \end{bmatrix}^{-1} \begin{bmatrix} r_k \\ 0 \end{bmatrix}, \tag{2.29}$$

and $x_k^s := K_{xr} r_k$ and $\Delta u_k^s := K_{ur} r_k$. Subsequently, one can define a state-space model in terms of deviation variables:

$$\begin{aligned} \tilde{x}_{k+1} &= A\tilde{x}_k + B\Delta\tilde{u}_k, \\ \tilde{y}_k &= C\tilde{x}_k, \end{aligned} \tag{2.30}$$

where the deviation variables are defined as $\tilde{x}_{k+1} = x_k - x_k^s$, $\Delta\tilde{u}_k = \Delta u_k - \Delta u_k^s$ and $\tilde{y}_k = y_k - r_k$. Thus, the offset-free form of the cost function of (2.25) is as follows:

$$\min_{\Delta\tilde{u}_k} J := \sum_{k=0}^{\infty} \left(\tilde{x}_k^T Q_x \tilde{x}_k + \Delta\tilde{u}_k^T R \Delta\tilde{u}_k \right), \tag{2.31}$$

and the solution of (2.31) is the offset-free control law (2.27).

2.5.2.2 Cost Function in Dual-Mode MPC

Given that the LQR problem (2.31) is intractable when constraints are presented, thus, the basic idea of the dual-mode MPC is formulate the constraint optimisation problem of the form (2.31) in a way that can be solved simply by the quadratic programming method.

In typical dual-mode MPC, the prediction horizon is divided into two modes: a transient mode with degree-of-freedom (d.o.f.) and a terminal mode with convergence. For example, in the transient mode, a sequence of inputs is optimised over the control horizon n_c with respect to the handling of constraints, whilst in terminal mode, the closed-loop dynamics are governed by the pre-determined control law, which typically, is an asymptotically stabilising state feedback controller K computed by solving an unconstrained linear quadratic regulator (LQR) problem. Typically, the offset-free predictions of input and state of the system (2.30) are as follows:

$$\Delta\tilde{u}_{i|k} = \begin{cases} \Delta\tilde{u}_{i|k}, & \forall i < n_c, \\ -K\tilde{x}_{i|k}, & \forall i \geq n_c, \end{cases} \tag{2.32a}$$

$$\tilde{x}_{i+1|k} = \begin{cases} A\tilde{x}_{i|k} + B\Delta\tilde{u}_{i|k}, & \forall i < n_c, \\ (A - BK)\tilde{x}_{i|k}, & \forall i \geq n_c, \end{cases} \tag{2.32b}$$

where $\Phi := A - BK$. Thus, the cost function of (2.31) is parametrised into two modes in the dual-mode MPC, that is defined as follows:

$$J_k = \underbrace{\sum_{i=0}^{n_c-1} \left(\tilde{x}_{i|k}^T Q_x \tilde{x}_{i|k} + \Delta\tilde{u}_{i|k}^T R \Delta\tilde{u}_{i|k} \right)}_{\text{Transient mode}} + \underbrace{\sum_{i=n_c}^{\infty} \left(\tilde{x}_{i|k}^T Q_x \tilde{x}_{i|k} + \Delta\tilde{u}_{i|k}^T R \Delta\tilde{u}_{i|k} \right)}_{\text{Terminal mode}}. \tag{2.33}$$

By substituting the predictions in (2.32) into (2.33), the cost function in the *terminal mode* can be simplified as follows:

$$\sum_{i=n_c}^{\infty} \left(\tilde{x}_{i|k}^T Q_x \tilde{x}_{i|k} + \Delta\tilde{u}_{i|k}^T R \Delta\tilde{u}_{i|k} \right), \tag{2.34a}$$

$$= \sum_{i=0}^{\infty} \left(\tilde{x}_{n_c|k}^T (\Phi^i)^T Q_x \Phi^i \tilde{x}_{n_c|k}^T + \tilde{x}_{n_c|k}^T (\Phi^i)^T K^T R K \Phi^i \tilde{x}_{n_c|k}^T \right), \tag{2.34b}$$

$$= \tilde{x}_{n_c|k}^T \underbrace{\sum_{i=0}^{\infty} \left((\Phi^i)^T \underbrace{(Q_x + K^T R K)}_{W} \Phi^i \right)}_{\Sigma} \tilde{x}_{n_c|k}^T, \tag{2.34c}$$

where Σ is the solution of a Lyapunov equation $\Phi^T \Sigma \Phi = \Sigma - W$, which can be easily solved using `dlyap.m` in Matlab®. Thus, the cost (2.33) becomes:

$$J_k = \sum_{i=0}^{n_c-1} \left(\tilde{x}_{i|k}^T Q_x \tilde{x}_{i|k} + \Delta\tilde{u}_{i|k}^T R \Delta\tilde{u}_{i|k} \right) + \tilde{x}_{n_c|k}^T \Sigma \tilde{x}_{n_c|k}^T, \tag{2.35a}$$

$$= \underset{\rightarrow k}{\tilde{x}}^T \bar{Q}_x \underset{\rightarrow k}{\tilde{x}} + \Delta \underset{\rightarrow k-1}{\tilde{u}}^T \bar{R} \Delta \underset{\rightarrow k-1}{\tilde{u}}, \tag{2.35b}$$

where $\bar{Q}_x \in \mathbb{R}^{n_c n_x \times n_c n_x}$ and $\bar{R} \in \mathbb{R}^{(n_c n_u - n_u) \times (n_c n_u - n_u)}$ are diagonal matrices, that are defined as follows:

$$\bar{Q}_x := \begin{bmatrix} Q_x & & \\ & \ddots & \\ & & Q_x \\ & & & \Sigma \end{bmatrix}, \quad \bar{R} := \begin{bmatrix} R & & \\ & \ddots & \\ & & R \end{bmatrix}. \tag{2.35c}$$

Notice that the dual-mode cost function (2.35b) has an equivalent form as the GPC cost (2.20a) if the terminal weight Σ is incorporated into the GPC cost (2.20a).

2.5.2.3 Constraints in Dual-Mode MPC

Constraint formulation in dual-mode MPC is divided into two modes as well. In the transient mode, it is expected the degree-of-freedom $\Delta \underset{\rightarrow k-1}{u}$ satisfied a set of linear equalities similar to (2.23), that is defined as follows:

$$\Delta u_{\min}\mathbf{1} \leq \Delta \underset{\rightarrow k-1}{u} \leq \Delta u_{\max}\mathbf{1}, \tag{2.36a}$$

$$x_{\min}\mathbf{1} \leq \underset{\rightarrow k}{x} \leq x_{\max}\mathbf{1}, \tag{2.36b}$$

$$y_{\min}\mathbf{1} \leq C\underset{\rightarrow k}{x} \leq y_{\max}\mathbf{1}. \tag{2.36c}$$

In the terminal mode, in order to ensure constraint satisfactions, possible violations in (2.23) must be checked over an infinite prediction horizon:

$$\begin{bmatrix} -K \\ K \\ I \\ -I \\ C \\ -C \end{bmatrix} \Phi^i x_{n_c|k} \leq \begin{bmatrix} u_{\max} \\ -u_{\min} \\ x_{\max} \\ -x_{\min} \\ y_{\max} \\ -y_{\min} \end{bmatrix}, \quad \forall i \geq 0, \tag{2.37}$$

which would appear to be computationally intractable. However, it is well known (Gilbert & Tan, 1991) that there exists a sufficiently large horizon $i = n_\infty$ where any additional linear equalities of (2.23) become redundant, assuming K is a stabilising feedback law and the constraints contain the steady-state within their interior. Therefore, the constraints (2.37) just need to be checked up to the horizon n_∞.

Collecting the cost (2.35b) and constraint (2.36) and (2.37), the dual-mode algorithm can be stated as follows:

Algorithm 2.5.2 (*Dual-mode MPC*) At every sample k,

1. Perform the optimisation

$$\arg\min_{\underset{\to k-1}{\Delta u}} \quad J_k := \underset{\to k}{\tilde{x}}^T \bar{Q}_x \underset{\to k}{\tilde{x}} + \Delta \underset{\to k-1}{\tilde{u}}^T \bar{R} \Delta \underset{\to k-1}{\tilde{u}}, \tag{2.38a}$$

$$\text{s.t.} \quad \underset{\to k}{\tilde{x}} = P_{xx}\tilde{x}_{0|k} + H_x \Delta \underset{\to k-1}{\tilde{u}}, \tag{2.38b}$$

$$(2.36), \tag{2.38c}$$

$$(2.37), \tag{2.38d}$$

   where $x_{0|k} = x_k$ is obtained from the plant at sample k.
2. Apply the first element of the control sequence $\Delta \underset{\to k-1}{u}$ to the plant, where $\Delta \underset{\to k-1}{u}$ is the solution of (2.38).

2.5.3 *Closed-Loop Paradigm*

One of the useful ways to set up the dual-mode MPC is the closed-loop paradigm, that is widely employed in the MPC and wind turbine control literature (e.g. Kumar & Stol, 2009; Rossiter, Kouvaritakis, & Rice, 1998; Spencer, Stol, Unsworth, Cater, & Norris, 2013). The beauty of the closed-loop paradigm approach is that the degrees-of-freedom is formulated as perturbations $c_k \in \mathbb{R}^{n_u}$ around the unconstrained optimal control law $\Delta\tilde{u}_k = -K\tilde{x}_k$, thus, one can gain insights into the impact of constraints by examining the magnitude of the perturbation c_k. The predictions (2.32) under the closed-loop paradigm become (e.g. Rossiter et al., 1998):

$$\Delta\tilde{u}_{i|k} = \begin{cases} -K\tilde{x}_{i|k} + c_{i|k}, & \forall i < n_c, \\ -K\tilde{x}_{i|k}, & \forall i \geq n_c, \end{cases} \tag{2.39}$$

$$\tilde{x}_{i|k} = \begin{cases} \Phi\tilde{x}_k + Bc_{i|k}, & \forall i < n_c, \\ \Phi\tilde{x}_{i|k}, & \forall i \geq n_c. \end{cases} \tag{2.40}$$

The cost (2.35b) function become:

$$J_k = \sum_{i=0}^{n_c-1} \left(\tilde{x}_{i|k}^T Q_x \tilde{x}_{i|k} + (-K\tilde{x}_{i|k} + c_{i|k})^T R (-K\tilde{x}_{i|k} + c_{i|k}) \right) + \tilde{x}_{n_c|k}^T \Sigma \tilde{x}_{n_c|k}, \tag{2.41a}$$

$$= \sum_{i=0}^{n_c-1} c_{i|k}^T W_c c_{i|k} = \underset{\to k-1}{c}^T \bar{W}_c \underset{\to k-1}{c}, \tag{2.41b}$$

where $W_c := \text{diag}(B^T \Sigma B + R) \in \mathbb{R}^{n_u n_c \times n_u n_c}$.

The derivation of the cost (2.41) is as follows:

$$\begin{aligned}
J_k = &\sum_{i=0}^{n_c-2} \left(\tilde{x}_{i|k}^T Q_x \tilde{x}_{i|k} + (-K\tilde{x}_{i|k} + c_{i|k})^T R(-K\tilde{x}_{i|k} + c_{i|k})\right) \\
&+ (-K\tilde{x}_{n_c-1|k} + c_{n_c-1|k})^T R(-K\tilde{x}_{n_c-1|k} + c_{n_c-1|k}) \\
&+ \tilde{x}_{n_c-1|k}^T Q\tilde{x}_{n_c-1|k} + (\Phi\tilde{x}_{n_c-1|k} + Bc_{n_c-1|k})^T \Sigma(\Phi\tilde{x}_{n_c-1|k} + Bc_{n_c-1|k}),
\end{aligned} \tag{2.42a}$$

$$= \sum_{i=0}^{n_c-2} \left(\tilde{x}_{i|k}^T Q_x \tilde{x}_{i|k} + (-K\tilde{x}_{i|k} + c_{i|k})^T R(-K\tilde{x}_{i|k} + c_{i|k})\right) \tag{2.42b}$$

$$\begin{aligned}
&+ \tilde{x}_{n_c-1|k}^T (Q + K^T RK + \Phi^T \Sigma\Phi)\tilde{x}_{n_c-1|k}^T \\
&+ c_{n_c-1|k}^T (B^T \Sigma B + R)c_{n_c-1|k} + 2\tilde{x}_{n_c-1|k}^T (\Phi^T \Sigma B - K^T R)c_{n_c-1|k}.
\end{aligned} \tag{2.42c}$$

Notice that $\Phi^T \Sigma B - K^T R = 0$ based on $\Phi = A - BK$ and $K = (B^T \Sigma B + R)^{-1}(B^T \Sigma A)$ is from the discrete algebraic Riccati equation of the infinite-horizon optimal control problem (2.25). In addition, $Q + K^T RK + \Phi^T \Sigma\Phi = \Phi$, which is the Lyapunov equation as shown earlier. Thus, (2.42c) eventually leads to:

$$J_k = \sum_{i=0}^{n_c-1} \left(c_{i|k}^T W_c c_{i|k}\right) + x_{0|k}^T \Sigma x_{0|k}, \tag{2.43}$$

where $x_{0|k}^T \Sigma x_{0|k}$ is independent of the degree-of-freedoms $c_{i|k}$, and thus can be omitted.

If the constraints (2.36) and (2.38) are expressed in terms of $\underset{\rightarrow k-1}{c}$, then the dual-mode MPC optimisation problem (2.38) can be formulated in terms of the perturbation, that minimises the cost function $J_k = \underset{\rightarrow k-1}{c}^T \bar{W}_c \underset{\rightarrow k-1}{c}$ subject to constraints.

To sum up, in *unconstrained* situation, minimising the cost function (2.41) simply yields $\underset{\rightarrow k-1}{c} = 0$, that implies the control law is the feedback control law $\Delta\tilde{u}_k = -K\tilde{x}_k$. When *constraints are active*, the magnitude of $\underset{\rightarrow k-1}{c}$ reveals how far the input is away from the unconstrained optimum. This offers important insight into how the dual-mode MPC handles the feed-forward information, that is discussed in Chap. 3.

2.6 Fatigue Load Assessment

As the blade pitch control offers a means of attenuating loads on turbine structures, it is necessary in this section to introduce the concept of fatigue loads.

Fatigue is the cumulative structural damage that occurs when a material is subjected to repetitive loads. The structural components of a wind turbine are often subjected to a sequence of repetitive loadings, and such loadings mainly arise from wind-shear, gravitation, harmonic effects of the rotor rotation, and so on (Barlas & van Kuik, 2010). The fatigue lifetime of turbine components is an important criterion for wind turbine design and reliable operations. Therefore, any load reductions on key components can directly translate to an extension on the lifetime of a wind turbine and also reductions in numbers of the costly maintenance and repair activities.

The key mechanical components of a turbine such as the blades and tower are subjected to loads that covers a wide range of frequencies and magnitudes. These repetitive loads can be quantified as the fatigue damage equivalent load (DEL). The DEL are typically characterised by the rain-flow counting method, that is a widely used approach (Niesłony, 2009; Sutherland, 1999) to estimate the fatigue damage of a turbine component, by counting the number of cycles and their corresponding amplitude on a load signal, which is defined as follows (e.g. Sutherland, 1999):

$$M_{\text{DEL}} = \Big(\sum_j \frac{\Delta M_j^m n_j}{n}\Big)^{\frac{1}{m}}, \tag{2.44}$$

where $n_j \in \mathbb{R}$ denotes the *number of cycles to the failure* associated with the *cycle load magnitude* ΔM_j, whilst $n \in \mathbb{R}$ represent the total number of cycles. $m \in \mathbb{R}$ is a material-specific parameter. This material-specific parameter is typically 4 for steel, and 10 for composite (Bossanyi, 2003b).

Notice that the DEL (2.44) is not of a quadratic form, that implies a cost function consisting of the DEL could not be solved by the standard quadratic programming methods. However, the DEL of a load signal is highly correlated to its variance or standard deviation. This proves useful because minimisation of the variance of a load signals has a similar effect on its DEL. And the variance itself is of a quadratic form, that can be easily minimised by quadratic programming (Castaignet et al., 2013; Simley, Dunne, Laks, & Pao, 2013). To illustrate the correlation is high between the DEL and variance, Fig. 2.10 shows the relationship between the DEL and standard deviation of a measurement signals of the flap-wise blade root bending moments obtained from the NREL 5-MW simulation turbine under various wind conditions.

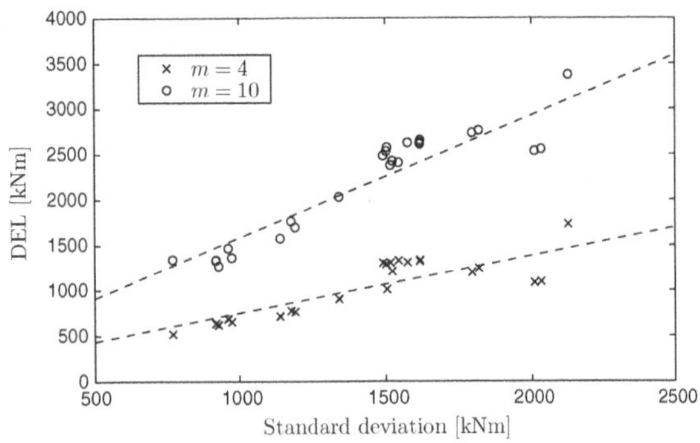

Fig. 2.10 Relationship of the damage equivalent load (DEL) and standard deviation. Data represents measurements of flap-wise blade root bending moment obtained from several simulations with material-specific parameter $m = 4$ and $m = 10$

Notice that two blade materials were examined, that they are steel ($m = 4$) and composite ($m = 10$). The dashed lines in Fig. 2.10 showed that the DEL and the standard deviation of the loads were in a proportional relationship. Moreover, the figure also reveals that the same load reductions in terms of the variance could translate to a higher fatigue reduction for materials with higher specific parameter m. In other words, the choice of materials affected the interpretation of the percentage reduction in fatigue loads. Hence, it might be fairer if load reduction strategies are accessed in terms of the variance or standard deviation. As a result, this thesis uses the standard deviation for assessing the load reduction for the blade pitch strategies.

2.7 Simulation Environment

In order to validate the performance of the controller and estimator design, this thesis employs a non-linear wind turbine simulation model, that is the FAST (Fatigue, Aerodynamics, Structures, and Turbulence) simulation package developed by the National Renewable Energy Laboratory (NREL) (Jonkman & Buhl Jr., 2005). The FAST code is open-source and widely used for wind turbine simulations in the academic community.

Compared to the linear models discussed in Sect. 2.4, the simulation turbine in the FAST is of much greater complexity, and with 24 degrees-of-freedom, including flap-wise and edge-wise blade modes, tower and shaft dynamics. Nonetheless, pitch actuator dynamics is not included in the simulation turbine. Thus, to limit the bandwidth of pitch actuator, this thesis places the linear actuator models from (2.15) in the high-fidelity simulation turbine. In terms of the fidelity of the wind data, the turbulent wind fields, that are employed by closed-loop simulations, are generated from TurbSim (Jonkman, 2009). The TurbSim code simulates a time series of wind data at points in a two-dimensional grid such that the sequence of grids marching towards the rotor at a constant speed specified by the mean wind speed and under the assumption of Taylor's frozen turbulence hypothesis (Taylor, 1938). The Taylor's frozen turbulence hypothesis assumes the spatial structure of the turbulent wind remains unchanged as they travel towards the wind turbine with the mean wind speed. The full-field three-dimensional wind data is characterised by mean wind speeds, turbulence settings and wind shear exponent, which are the input from the user. Subsequently, the AeroDyn (Laino & Hansen, 2002) code inside FAST calculates how the wind inflow data from TurbSim interact with the high degree-of-freedoms structural models. These structural models inside FAST are built based on the blade element momentum theory including the effect of dynamic stall and tower shadow.

The National Renewable Energy Laboratory (NREL) also published a benchmark wind turbine model, a 5 MW reference turbine (Jonkman et al., 2009). This fictitious NREL 5 MW turbine model was developed based on a number of commercial wind turbines, for example, Repower 5 MW. Given that the benchmark turbine model is available to the public without any commercial restriction, the NREL 5 MW reference

Table 2.1 Simulation turbine parameters

Rating	5 MW
Rotor configuration	Upwind, 3 blades
Rotor diameter	126 m
Hub height	90 m
Rated rotor speed	12.1 rpm ($\approx$0.2 Hz)
Cut-in, rated, cut-out wind speed	3, 11.4, 25 ms^{-1}

turbine has been widely adopted in many wind turbine control studies. Table 2.1 summarises some key parameters of the NREL 5 MW simulation turbine. Full details regarding this turbine can be found in Jonkman et al. (2009).

2.8 Baseline Controller Designs

The section provides some background of the baseline robust CPC and IPC controller. The reason why the controller design is presented in this early chapter is because these baseline controllers are employed in most of the coming chapters.

2.8.1 *Collective Pitch Controller*

Most of the modern wind turbine are equipped with a collective pitch controller for regulating the rotor speed as discussed in Sect. 1.2. The CPC controller provides collective blade pitch signals in response to the rotor speed deviation in the above-rated wind conditions. Thus, consider the rotor dynamics mapping the collective pitch angle $\bar{\theta}$ to the rotor speed ω including the actuator and filter dynamics as follows:

$$G_{\mathrm{cpc}}(s) := G_{\mathrm{a}}(s)G_{\mathrm{r}}(s)G_{\mathrm{lp}}(s), \tag{2.45a}$$

where the rotor dynamics $G_{\mathrm{r}} \in \mathcal{R}$, the actuator and low-pass filter dynamics $G_{\mathrm{a}}, G_{\mathrm{lp}} \in \mathcal{R}$ are provided in Sect. 2.4 and they are defined as follows:

$$G_{\mathrm{r}}(s) := \frac{d\tau_{\mathrm{a}}}{d\theta}\frac{1}{J_{\mathrm{r}}s + D_{\mathrm{r}}}, \tag{2.45b}$$

$$G_{\mathrm{a}}(s) := \frac{1}{\tau_{\mathrm{act}}s + 1}, \tag{2.45c}$$

$$G_{\mathrm{lp}}(s) := \frac{2\pi f_{\mathrm{lp}}}{s + 2\pi f_{\mathrm{lp}}}, \tag{2.45d}$$

where the parameter is provided in Table 2.2.

Table 2.2 Model parameters of $G_{\mathrm{cpc}}(s)$ and $K_{\mathrm{cpc}}(s)$

Parameters	Values	Units	Parameters	Values	Units
$\frac{d\tau_{\mathrm{a}}}{d\theta}$	-8.12×10^6	Nm deg^{-1}	J_{r}	3.88×10^7	kg m^2
D_{r}	9.68×10^6	–	τ_{act}	0.11	s
K_{p}	6.84	–	K_{i}	2.45	–
f_{lp}	0.15	Hz			

Remark 2.3 Notice that the transfer function representation of the turbine models is used in this section because the feedback controller is designed in the frequency-domain to ensure the gain of the loop-shape is only high at the targeted loads, explained as follows.

Typically, and for the reasons of implementation simplicity, the CPC and IPC are designed separately and the proportional-integral (PI) controllers are widely adopted for the CPC in the industry (Burton et al., 2011). This is because for the single-input-single-output problem, the PI controller is a simple but effective form of control strategy. As the name suggests, the control action of a PI controller is computed based on two gains: the proportional gain $K_{\mathrm{p}} \in \mathbb{R}$, that is proportional to the error between the desired and actual rotor speed and the integral gain $K_{\mathrm{i}} \in \mathbb{R}$, that is proportional to the integral of the error and ensures the error tends to zero (Aström & Murray, 2010). A typical PI controller is defined as follows:

$$K_{\mathrm{pi}}(s) = K_{\mathrm{p}} + \frac{K_{\mathrm{i}}}{s}. \tag{2.46}$$

The CPC controller is designed based on the $G_{\mathrm{cpc}} \in \mathcal{R}$ from (2.45), whose parameters are listed in Table 2.2, and which has the following transfer function:

$$G_{\mathrm{cpc}}(s) := \frac{-1.789}{s^3 + 10.28s^2 + 11.07s + 2.13}. \tag{2.47}$$

Notice that as mentioned in Sect. 2.4, the controller needs to be robust to the model uncertainties, since the linear model is obtained at one operating wind condition $v = 18\,\mathrm{ms}^{-1}$. Thus, while tuning the proportional and integral gain, one needs to ensure the resultant closed-loop of the rotor dynamics has a good gain and phase margin. In addition, the couplings between the CPC and IPC need to be taken into account. As discussed in Sect. 1.2, the IPC typically responds to the blade loads at the harmonics of the blade rotational frequency (e.g. 0.2 Hz). Thus, based on the rotor model (2.47), the CPC is tuned as follows:

$$K_{\mathrm{cpc}}(s) = -\frac{10.75s + 3.85}{3.14s}, \tag{2.48}$$

where the values of the proportional gain K_{p} and integral gain K_{i} are listed in Table 2.2. The loop gain of $G_{\mathrm{r}}(s)K_{\mathrm{cpc}}(s)$ is low (i.e. below 0 dB at frequencies beyond

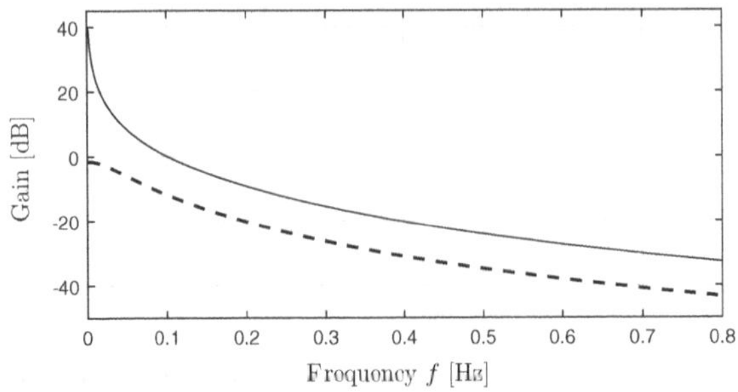

Fig. 2.11 Loop gain of $G_{\text{cpc}}(s)$ (dashed line) and $G_{\text{cpc}}(s)K_{\text{cpc}}(s)$ (solid line)

0.1 Hz as shown in Fig. 2.11. The compensated system $G_{\text{cpc}}(s)K_{\text{cpc}}(s)$ has an infinity gain margin and 82.9 deg phase margin.

The performance of the CPC is illustrated in Fig. 2.12, where the simulation is conducted in the NREL 5 MW reference turbine (Jonkman et al., 2009). As shown in Fig. 2.12a, the rotor speed deviation is attenuated by the CPC controller (2.48). And Fig. 2.12b reveals the pitch activity of the CPC.

2.8.2 *Individual Pitch Controller*

Individual pitch control strategies are one of the promising load attenuation techniques. Typically, most of the load fluctuations on the blade are of a periodic nature and some of the stochastic components. As the blade rotates in a turbulent wind field, the spatially uneven wind field across the rotor plane causes the blade to experience loads at 1p (once-per-revolution) or higher harmonics frequencies, for example, 2p (twice-per-revolution) frequency. The major contributors to the blade loads are the horizontal and vertical wind shear, gravitational forces, tower shadow, turbulence and yaw misalignment (Barlas & van Kuik, 2010). Thus, this explains why the loads on the rotating blades are predominately concentrated at harmonics of the blade rotational frequency. To visualise, consider a wind turbine simulation with a CPC and no IPC. Figure 2.13 shows the frequency spectrum of the typical flap-wise blade root bending moments. Given that the blade rotational frequency, or the rotor speed, is around 12.1 rpm $\approx$ 0.2 Hz, thus, the majority of the blade loads exists at 0.2 and 0.4 Hz, as demonstrated in Fig. 2.13. Notice that there exist blade loads around 0 Hz and these slow-varying loads should not be attenuated, otherwise, the IPC interferes with the CPC.

In addition, these loads on the rotating blades cause fluctuations not only on the blades, but also on the non-rotating turbine structure such as the main bearing. For a three-bladed turbine, in general, the rotating blade loads at integer multiples of the 1p blade frequency are transferred into non-rotating loads at adjacent harmonics at

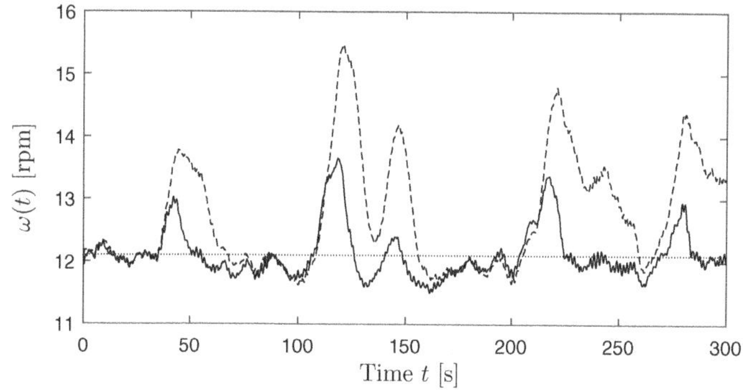

(A) The open-loop rotor speed response (dashed line) and closed-loop response (solid line). The rated rotor speed is denoted as dotted line.

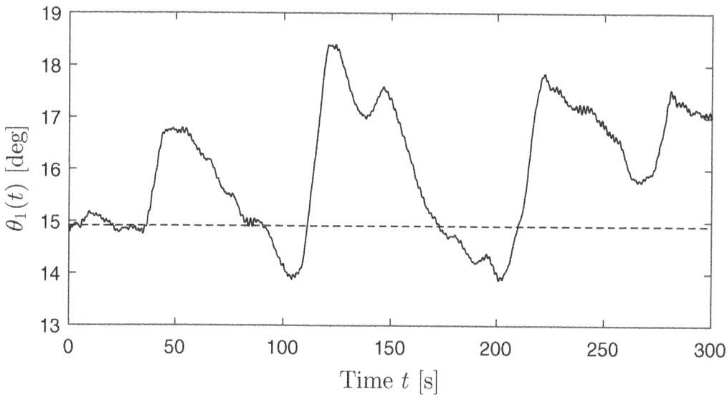

(B) The pitch angle of blade 1 of the CPC controller (solid line) and the open-loop (dashed line). Similar results are obtained for the remaining blades.

Fig. 2.12 Performance of the CPC controller to a time-varying wind speed disturbance in the above-rated wind conditions

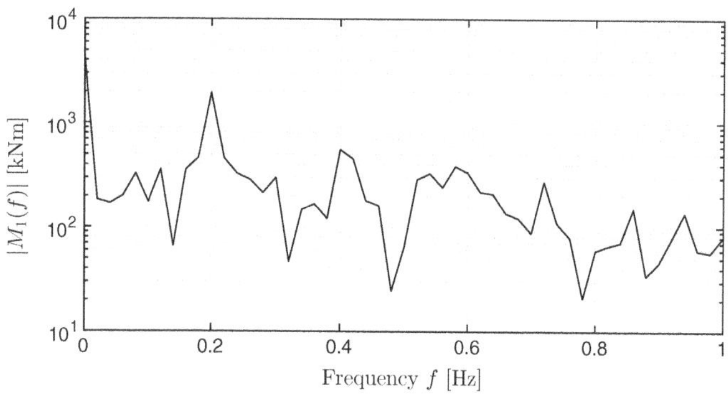

Fig. 2.13 Frequency spectrum of the flap-wise blade root bending moments of blade 1

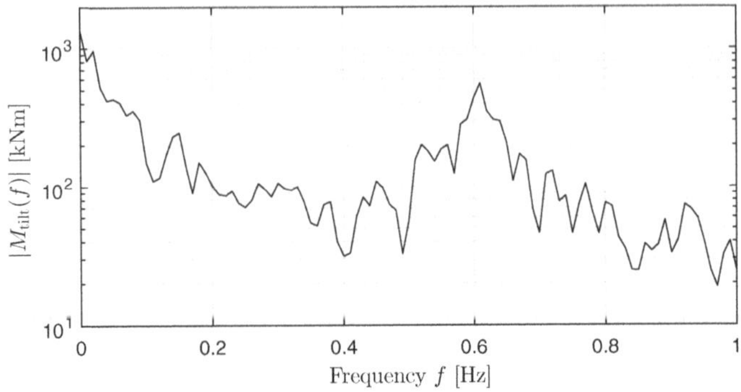

(A) Frequency spectrum of the main bearing tilt bending moments.

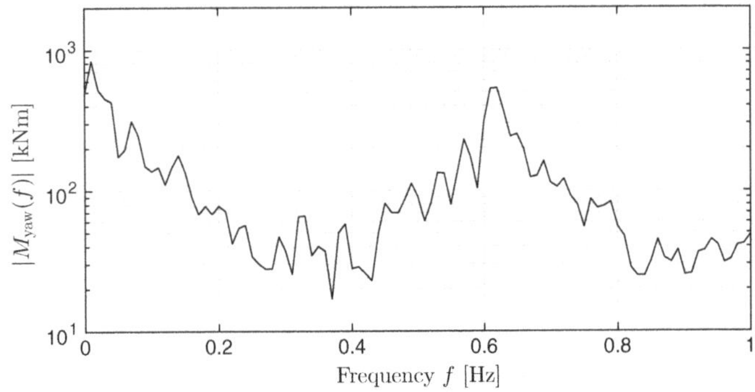

(B) Frequency spectrum of the main bearing yaw bending moments.

Fig. 2.14 Illustrations of the loads on the main bearing

frequency multiples of 3p. For example, 1p blade rotating blade loads are mapped into static (0p) loads on the tilt and yaw reference frame, whilst 3p non-rotating structural loads are split into 2p and 4p blade rotating loads. To visualise these loads on the fixed turbine structures, Fig. 2.14 illustrates the main bearing tilt and yaw moments. It shows clearly that the loads are concentrated at the 0p and 3p frequencies on both tilt and yaw moments of the main bearing. The idea of the frequency-shifting effect on loads becomes clearer in Chap. 4, where Coleman transform projects loads from a rotating coordinate frame into a non-rotating reference frame.

To ensure the IPC is decoupled from the existing CPC, a robust $\mathcal{H}_\infty$ loop-shaping design method is employed.

2.8.2.1 $\mathcal{H}_\infty$ Loop-Shaping Design Method

In the literature, there exists a substantial amount of mature theories on linear control design. Nevertheless, it is crucial to choose one that is suitable to the IPC problem. As discussed in Sect. 2.4, the linear turbine model neglects many of the dynamics of an actual wind turbine. Thus, it is reasonable to employ a robust control design technique and such a technique is able to construct controllers that are insensitive to

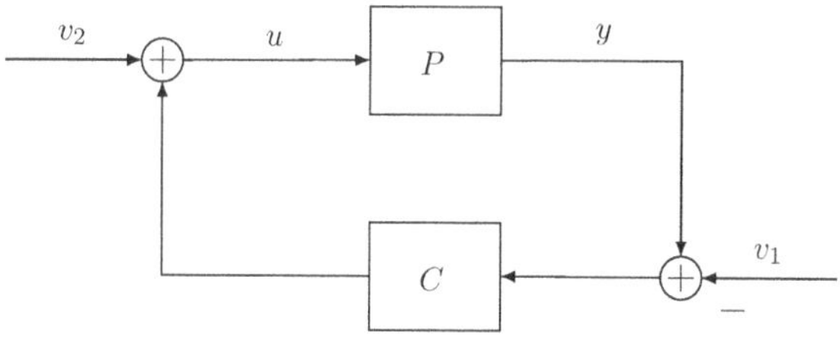

Fig. 2.15 Standard feedback interconnection between plant P and controller C. The signals u and y denote the plant input and measured output, respectively, whilst v_1 and v_2 represent exogenous disturbances

unmodelled dynamics. In addition, it is desirable that the load reduction performance can be specified in the frequency-domain, since the blade loads mainly occur at the harmonics of the blade rotational frequency. These requirements naturally suggest that the use of $\mathcal{H}_\infty$ loop-shaping (Vinnicombe, 2000). Before discussions on the design procedure, it is necessary to introduce the concept of the robust stability margin.

Robust stability margin

The standard feedback interconnection $[P, C]$ of plant $P \in \mathcal{R}$ and controller $C \in \mathcal{R}$ is shown in Fig. 2.15, from which the following closed-loop system is defined:

$$\begin{bmatrix} y \\ u \end{bmatrix} = \underbrace{\begin{bmatrix} P \\ I \end{bmatrix} (I - CP)^{-1} \begin{bmatrix} -C & I \end{bmatrix}}_{H(P,C)} \begin{bmatrix} v_1 \\ v_2 \end{bmatrix}, \tag{2.49}$$

where $H(P, C) \in \mathcal{R}$ provided $[P, C]$ is well posed, and I is an identity matrix of compatible dimension. The robust stability margin $b(P, C) \in \mathbb{R}$ of $[P, C]$ is defined as follows:

$$b(P, C) := \begin{cases} \|H(P, C)\|_\infty^{-1} & \text{if } H(P, C) \in \mathcal{RH}_\infty \\ 0 & \text{otherwise.} \end{cases} \tag{2.50}$$

where $|| \cdot ||_\infty$ is the $\mathcal{H}_\infty$ norm.

The robust stability margin is briefly a scalar quantity that generalises the gain and phase margins of a multivariable system. And the higher values of the robust stability margin give rise to the closed-loop system that is more robust to parametric uncertainties and unmodelled high-frequency dynamics. The objective of $\mathcal{H}_\infty$ loop-shaping method is briefly to find a stabilising controller $C \in \mathcal{R}$ that maximises the robust stability margin $b(P, C)$. In the next section, the design procedure is demonstrated.

Design procedure

The design of $\mathcal{H}_\infty$ loop-shaping controllers can be summarised into two steps as follows:

1. Considering $P(s)$ is the plant model, design a pre-compensator $W(s)$ to shape the singular values of the shaped system $P_s(s) := P(s)W(s)$. Typically, the shaped system $P_s(s)$ needs to have high loop gain at the targeted load frequencies and low gain at high frequencies for good sensor noise rejection.
2. Synthesise the controller $C_\infty(s)$ that achieves maximum robust stability margin of the shaped plant $P_s(s)$. Subsequently, the controller $C_\infty(s)$ is then combined with the pre-compensator $W(s)$ to form the actual loop-shaping controller $C(s) := W(s)C_\infty(s)$.

Notice that the the loop-shaping controller $C(s)$ can be easily synthesised by software package such as `ncfsyn.m` in Matlab®. Hence, the mathematical detail how to compute the controller C is not included in the thesis, the reader is referred to (Vinnicombe, 2000).

2.8.2.2 Design of the IPC

After the introduction of the characteristics of the blade loads and $\mathcal{H}_\infty$ loop-shaping method, let's consider the design of the IPC. Figure 2.16 depicts a typical blade pitch control system architecture for the above-rated wind conditions. The CPC regulates the rotor speed $\omega(t)$ by adjusting the collective pitch angle $\bar{\theta}(t)$. To isolate the actions of the IPC from that of the CPC, it is convenient to define the pitch angles and blade moments as follows:

$$\begin{bmatrix} \theta_1(t) \\ \theta_2(t) \\ \theta_3(t) \end{bmatrix} := \begin{bmatrix} \bar{\theta}(t) + \tilde{\theta}_1(t) \\ \bar{\theta}(t) + \tilde{\theta}_2(t) \\ \bar{\theta}(t) + \tilde{\theta}_3(t) \end{bmatrix}, \quad \begin{bmatrix} M_1(t) \\ M_2(t) \\ M_3(t) \end{bmatrix} := \begin{bmatrix} \bar{M}(t) + \tilde{M}_1(t) \\ \bar{M}(t) + \tilde{M}_2(t) \\ \bar{M}(t) + \tilde{M}_3(t) \end{bmatrix}, \tag{2.51}$$

where $\tilde{\theta}_{1,2,3}(t)$ represent the perturbations in the blade pitch angle demand from the collective pitch signal, whilst $\tilde{M}_{1,2,3}(t)$ are the perturbations in the flap-wise blade bending moments, obtained by filtering out the mean moment $\bar{M}(t)$ from the measurements $M_{1,2,3}(t)$. This filtering is important in order to help decouple the IPC from the CPC. For each blade, the relationship between perturbation input $\tilde{\theta}_i$ and output $\tilde{M}_i$, for $i \in \{1, 2, 3\}$ can be modelled by a transfer function $G \in \mathcal{R}$, obtaining by linearising the turbine dynamics around the rated rotor speed ω_0. A typical blade transfer function that maps the perturbation in the blade pitch angle $\tilde{\theta}_{1,2,3}$ to the perturbations in the flap-wise blade bending moments $\tilde{M}_{1,2,3}$, as used by Lu et al. (2015) for example, is as follows:

$$G(s) := G_a(s)G_b(s)G_{bp}(s), \tag{2.52a}$$

where G_a, $G_b \in \mathcal{R}$ describe the dynamics of the pitch actuator and the blade, respectively, whilst $G_{bp} \in \mathcal{R}$ is a band-pass filter that is included in order to remove the low frequency component of the flap-wise blade root bending moment signal from

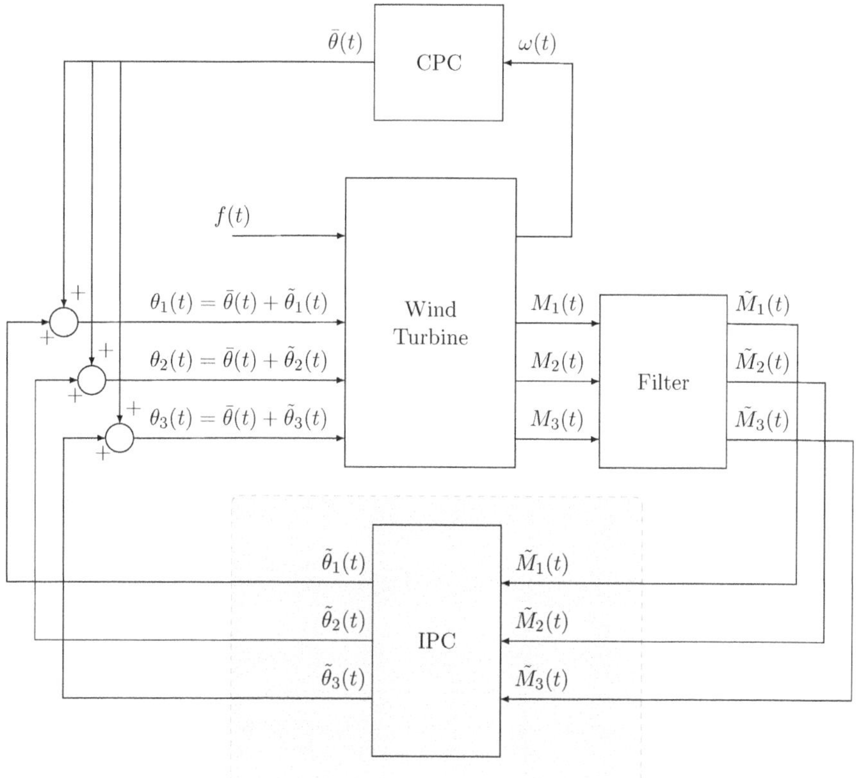

Fig. 2.16 System architecture of a wind turbine, combining collective pitch control (CPC) and individual pitch control (IPC). The CPC regulates rotor speed while the IPC (shaded) attenuates perturbations in the flap-wise root bending moments on each blade. Additional inputs to the turbine such as wind loading and generator torque are accounted for in the term $f(t)$

strain-gauge sensors, as well as high frequency noise. These transfer functions are obtained from Sect. 2.4 and they are defined as follows:

$$G_{\mathrm{a}}(s) := \frac{1}{1+\tau_{\mathrm{act}}s}, \tag{2.52b}$$

$$G_{\mathrm{b}}(s) := \frac{dM_{\mathrm{flap}}}{d\theta}\frac{(2\pi f_{\mathrm{b}})^2}{s^2 + D_{\mathrm{b}}2\pi f_{\mathrm{b}}s + (2\pi f_{\mathrm{b}})^2}, \tag{2.52c}$$

$$G_{\mathrm{bp}}(s) := \frac{2\pi f_{\mathrm{h}}s}{s^2 + 2\pi(f_{\mathrm{h}} + f_{\mathrm{l}})s + 4\pi^2 f_{\mathrm{h}} f_{\mathrm{l}}}, \tag{2.52d}$$

where $\tau_{\mathrm{act}} \in \mathbb{R}$ is the pitch actuator time constant, $\frac{dM_{\mathrm{flap}}}{d\theta} \in \mathbb{R}$ represents the change in blade flap-wise bending moment with respect to pitch angle, $f_{\mathrm{b}} \in \mathbb{R}$ is the natural frequency of the blade's first flap-wise mode and $D_{\mathrm{b}} \in \mathbb{R}$ is its aerodynamic damping ratio, while $f_h, f_l \in \mathbb{R}$ are the high and low corner frequencies of the bandpass filter, respectively. The basic individual pitch control problem is shown in Fig. 2.17 and is based upon the following three-blade model:

$$\begin{bmatrix} \tilde{M}_1(s) \\ \tilde{M}_2(s) \\ \tilde{M}_3(s) \end{bmatrix} = \underbrace{\begin{bmatrix} G(s) & 0 & 0 \\ 0 & G(s) & 0 \\ 0 & 0 & G(s) \end{bmatrix}}_{P(s)} \begin{bmatrix} \tilde{\theta}_1(s) \\ \tilde{\theta}_2(s) \\ \tilde{\theta}_3(s) \end{bmatrix}. \tag{2.53}$$

In the interests of simplicity, the influence of the fixed turbine structural dynamics is not included, but if required, these could be represented as additive disturbances on the bending moment channels.

The design of the IPC controller $K_{\text{ipc}} \in \mathcal{R}$ is based upon the basic blade model G (2.52a), whose parameters are listed in Table 2.3, and which has the following transfer function:

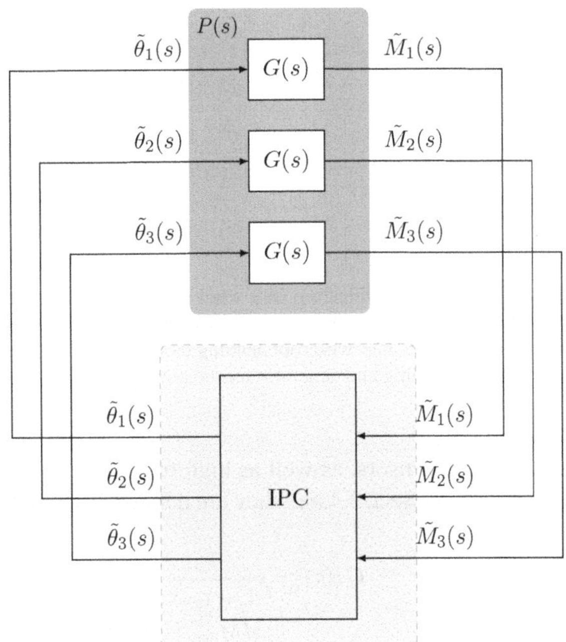

Fig. 2.17 Basic system architecture for IPC analysis and design

Table 2.3 Model parameters of $G(s)$

Parameters	Values	Units	Parameters	Values	Units
τ_{act}	0.11	s	$\frac{dM_{\text{flap}}}{d\theta}$	1.50×10^6	Nm deg^{-1}
f_b	0.70	Hz	D_b	0.47	
f_h	0.80	Hz	f_l	0.014	Hz

$$G(s) = \frac{1.45 \times 10^8 s}{0.11s^5 + 2.02s^4 + 13.84s^3 + 52.25s^2 + 101.50s + 8.54}. \tag{2.54}$$

The IPC controller needs to attenuate blade loads specifically at the 1p, 2p and 4p frequencies (0.2, 0.4 and 0.8 Hz respectively). Thus, it is convenient to use the robust $\mathcal{H}_\infty$ loop-shaping method to construct the IPC. To do so, the pre-compensator $W_{\text{ipc}}(s) \in \mathcal{R}$ is designed, such that the shaped plant $G(s)W_{\text{ipc}}(s)$ has high gain at the targeted frequencies, as follows:

$$W_{\text{ipc}}(s) := \alpha_{\text{ipc}} W_{\text{hp}}(s) W_{1\text{p}}(s) W_{2\text{p}}(s) W_{4\text{p}}(s), \tag{2.55a}$$

where $\alpha_{\text{ipc}} \in \mathbb{R}$ is a tuning variable and W_{hp}, $W_{1\text{p}}$, $W_{2\text{p}}$, $W_{4\text{p}} \in \mathcal{R}$ are the high-pass filter, 1p, 2p and 4p inverse notch filters, which are defined as follows:

$$W_{\text{hp}}(s) := \frac{\tau_{\text{hf}} s}{\tau_{\text{hf}} s + 1}, \tag{2.55b}$$

$$W_{1\text{p}}(s) := \frac{s^2 + 2D_{1\text{p}_1}\omega_{1\text{p}} s + \omega_{1\text{p}}^2}{s^2 + 2D_{1\text{p}_2}\omega_{1\text{p}} s + \omega_{1\text{p}}^2}, \tag{2.55c}$$

$$W_{2\text{p}}(s) := \frac{s^2 + 2D_{2\text{p}_1}\omega_{2\text{p}} s + \omega_{2\text{p}}^2}{s^2 + 2D_{2\text{p}_2}\omega_{1\text{p}} s + \omega_{2\text{p}}^2}, \tag{2.55d}$$

$$W_{4\text{p}}(s) := \frac{s^2 + 2D_{4\text{p}_1}\omega_{4\text{p}} s + \omega_{4\text{p}}^2}{s^2 + 2D_{4\text{p}_2}\omega_{4\text{p}} s + \omega_{4\text{p}}^2}. \tag{2.55e}$$

The parameters of this pre-compensator $W_{\text{ipc}}(s)$ are listed in Table 2.4. Subsequently, based upon the shaped plant $G(s)W_{\text{ipc}}(s)$, the loop-shaping controller $K_{\text{ipc}}(s) \in \mathcal{R}$, that maps the perturbation in the flap-wise blade root bending moment $\tilde{M}_{1,2,3}$ to the perturbation in the pitch angle $\tilde{\theta}_{1,2,3}$, is synthesised as follows:

Table 2.4 Model parameters of $K_{\text{ipc}}(s)$

Parameters	Values	Units	Parameters	Values	Units
α_{ipc}	0.45×10^{-3}	–	τ_{hf}	0.15	s
$\omega_{1\text{p}}$	1.27	rads^{-1}	$D_{1\text{p}_1}$	1.5	–
$D_{1\text{p}_2}$	0.01	–	$\omega_{2\text{p}}$	2.53	rads^{-1}
$D_{2\text{p}_1}$	1	–	$D_{2\text{p}_2}$	0.03	–
$\omega_{4\text{p}}$	5.07	rads^{-1}	$D_{4\text{p}_1}$	0.3	–
$D_{4\text{p}_2}$	0.03	–			

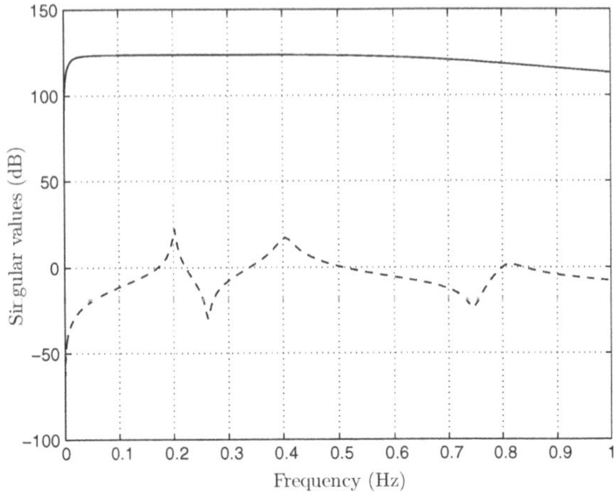

Fig. 2.18 Maximum singular value plots of the wind turbine model $\bar{\sigma}(G(s))$ (–) and compensated system $\bar{\sigma}(GK_{\text{ipc}}(s))$ (- -)

$$K_{\text{ipc}}(s) = 10^{-6} \times \left[\frac{\begin{array}{c} -s^9 - 5.6s^8 - 49.3s^7 - 174.0s^6 - 639.5s^5 \\ - 1236s^4 - 900s^3 - 2170s^2 + 1380s - 98 \end{array}}{\begin{array}{c} s^9 + 9.4s^8 + 87.2s^7 + 353.2s^6 + 1955s^5 \\ + 3031s^4 + 1.1 \times 10^4 s^3 + 7662s^2 + 1.3 \times 10^4 s + 5663 \end{array}} \right], \tag{2.56}$$

where the robust stability margin $b(GK_{\text{ipc}}, 1)$ is 0.58 and the gain and phase margins of the compensated system $G(s)K_{\text{ipc}}(s)$ are 10.3 dB and 52 deg. The gain of the compensated system $G(s)K_{\text{ipc}}(s)$ is illustrated in Fig. 2.18. Notice that the baseline IPC (2.56) is the simplest form of the IPCs, known as single-blade control (SBC). The other classes of IPCs involve coordinate transformations. More detail of various classes of IPCs is provided in Chaps. 3 and 4.

To examine the performance of the IPC controller (2.56), simulations were conducted under a time-varying wind disturbance upon the NREL 5 MW turbine (Jonkman et al., 2009). Comparison is made between the CPC controller, that is adopted from (2.48) as the baseline controller, and the IPC controller (2.56) on top of the baseline. Figure 2.19a illustrates the flap-wise blade root bending moments. The IPC controller attenuated the blade loads effectively compared to the simulation result with only the CPC controller. This is expected because the CPC and IPC system architecture requires more pitch activity as shown in Fig. 2.19b.

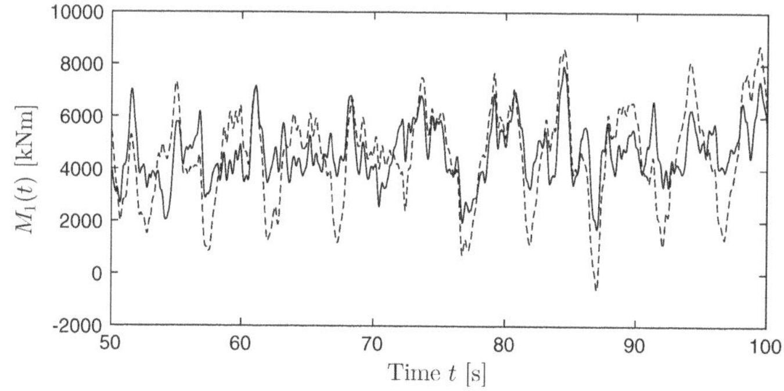

(A) The flap-wise blade root bending moment of the blade 1.

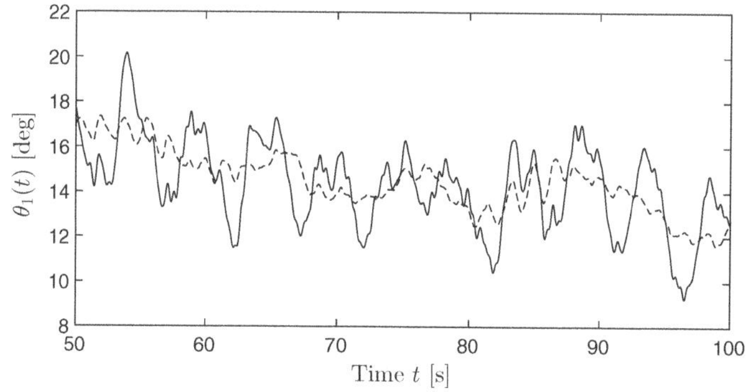

(B) The pitch angles of blade 1.

Fig. 2.19 Performance of the CPC (dased line) and with the IPC (solid line). Similar results are obtained for the remaining blades

2.9 Summary

This chapter presented background information regarding the wind turbine and load reduction control. The basic operations of a typical wind turbine are explained and the control objectives of wind turbines based on the modes of operations were discussed in Sects. 2.2 and 2.3. In addition, the detail of modelling was provided in Sect. 2.4. This included the linearisation of the aerodynamic forces, the dynamics of the rotor, tower and blade, and also the models of the actuator and filters. Section 2.6 explained why this thesis uses the standard deviation as an indicator of the fatigue damage. In Sect. 2.7, the detail of the simulation environment was discussed. Finally, Sect. 2.8 presented the design of the baseline CPC and IPC controllers, that is employed in most of the coming chapters.

References

Arnett, E. B., Huso, M. M., Schirmacher, M. R., & Hayes, J. P. (2011). Altering turbine speed reduces bat mortality at wind-energy facilities. *Frontiers in Ecology and the Environment*, *9*(4), 209–214.

Aström, K. J., & Murray, R. M. (2010). *Feedback systems: An introduction for scientists and engineers*. Princeton: Princeton University Press.

Barlas, T., & van Kuik, G. (2010). Review of state of the art in smart rotor control research for wind turbines. *Progress in Aerospace Sciences*, *46*(1), 1–27.

Bellman, R. (1957). *Dynamic programming*. Princeton: Princeton University Press.

Betz, A. (1966). *Introduction to the theory of flow machines (Vol. 1) (No. 1)*. Oxford: Permagon Press.

Bossanyi, E. A. (2000). The design of closed loop controllers for wind turbines. *Wind Energy*, *3*(3), 149–163.

Bossanyi, E. A. (2003a). Individual blade pitch control for load reduction. *Wind Energy*, *6*(2), 119–128.

Bossanyi, E. A. (2003b). Wind turbine control for load reduction. *Wind Energy*, *6*(3), 229–244.

Burton, T., Jenkins, N., Sharpe, D., & Bossanyi, E. (2011). *Wind energy handbook*. Chichester, UK: Wiley.

Castaignet, D., Barlas, T., & Buhl, T. (2013). Full scale test of trailing edge flaps on a Vestas V27 wind turbine: Active load reduction and system identification. *Wind Energy*, 2–6.

Chiang, M.-H. (2011). A novel pitch control system for a wind turbine driven by a variable-speed pump-controlled hydraulic servo system. *Mechatronics*, *21*(4), 753–761.

Clarke, D., Mohtadi, C., & Tuffs, P. (1987a). Generalized predictive control—Part II extensions and interpretations. *Automatica*, *23*(2), 149–160.

Clarke, D., Mohtadi, C., & Tuffs, P. (1987b). Generalized predictive control—Part I. The basic algorithm. *Automatica*, *23*(2), 137–148.

Clarke, D., & Scattolini, R. (1991). Constrained receding-horizon predictive control. *IEE Proceedings D Control Theory and Applications*, *138*(4), 347.

Dunne, F., Pao, L. Y., Wright, A. D., Jonkman, B., & Kelley, N. (2011). Adding feedforward blade pitch control to standard feedback controllers for load mitigation in wind turbines. *Mechatronics*, *21*, 682–690.

Geyler, M., & Caselitz, P. (2008). Robust multivariable pitch control design for load reduction on large wind turbines. *Journal of Solar Energy Engineering*, *130*(3), 031014.

Gilbert, E., & Tan, K. (1991). Linear systems with state and control constraints: The theory and application of maximal output admissible sets. *IEEE Transactions on Automatic Control*, *36*(9), 1008–1020.

Henriksen, L., Hansen, M., & Poulsen, N. (2012). Wind turbine control with constraint handling: A model predictive control approach. *IET Control Theory & Applications*, *6*(11), 1722.

Jonkman, B. (2009). *TurbSim user's guide* (Technical Report). National Renewable Energy Laboratory (NREL).

Jonkman, J., & Buhl Jr, M. (2005). *FAST user's guide* (Technical Report). National Renewable Energy Laboratory (NREL).

Jonkman, J., Butterfield, S., Musial, W., & Scott, G. (2009). *Definition of a 5-MW reference wind turbine for offshore system development* (Technical Report). Golden, CO: National Renewable Energy Laboratory (NREL).

Kouvaritakis, B., Rossiter, J. A., & Schuurmans, J. (2000). Efficient robust predictive control. *IEEE Transactions on Automatic Control*, *45*(8), 1545–1549.

Kumar, A., & Stol, K. (2009). Scheduled model predictive control of a wind turbine. In *Proceedings of AIAA/ASME Wind Energy Symposium*.

Laino, D. J., & Hansen, A. C. (2002). *AeroDyn user guide* (Technical Report). National Renewable Energy Laboratory (NREL).

Laks, J., Pao, L., & Wright, A. (2009). Control of wind turbines: Past, present, and future. In *2009 American Control Conference* (pp. 2096–2103). New York: IEEE.
Lu, Q., Bowyer, R., & Jones, B. (2015). Analysis and design of Coleman transformbased individual pitch controllers for wind-turbine load reduction. *Wind Energy*, *18*(8), 1451–1468.
Mirzaei, M., Henriksen, L. C., Poulsen, N. K., Niemann, H. H., & Hansen, M. H. (2012). Individual pitch control using LIDAR measurements. In *Proceedings of IEEE International Conference on Control Applications* (pp. 1646–1651).
Mosca, E., & Zhang, J. (1992). Stable redesign of predictive control. *Automatica*, *28*(6), 1229–1233.
Niesłony, A. (2009). Determination of fragments of multiaxial service loading strongly in fluencing the fatigue of machine components. *Mechanical Systems and Signal Processing*, *23*(8), 2712–2721.
Pannocchia, G., & Rawlings, J. B. (2003). Disturbance models for offset-free model predictive control. *AIChE Journal*, *49*(2), 426–437.
Pao, L., & Johnson, K. (2009). A tutorial on the dynamics and control of wind turbines and wind farms. In *Proceedings of ACC*.
Rawlings, J. B., & Muske, K. R. (1993). The stability of constrained receding horizon bibliography 191 control. *IEEE Transactions on Automatic Control*, *38*(10), 1512–1516.
Rossiter, J. A. (2003). *Model-based predictive control: A practical approach*. Boca Raton, FL: CRC Press.
Rossiter, J. A., Kouvaritakis, B., & Rice, M. J. (1998). A numerically robust state-space approach to stable-predictive control strategies. *Automatica*, *34*(1), 65–73.
Schlipf, D., Sandner, F., Raach, S., Matha, D., & Cheng, P. (2013). Nonlinear model predictive control of floating wind turbines. In *Proceedings of International Offshore and Polar Engineering* (Vol. 9, pp. 440–446).
Selvam, K., Kanev, S., van Wingerden, J. W., van Engelen, T., & Verhaegen, M. (2009). Feedback-feedforward individual pitch control for wind turbine load reduction. *International Journal of Robust and Nonlinear Control*, *19*(1), 72–91.
Simley, E., Dunne, F., Laks, J., & Pao, L. Y. (2013). *Lidars and wind turbine control Part 2* (Technical Report). DTU Wind Energy.
Soliman, M., Malik, O., & Westwick, D. (2010). Multiple model MIMO predictive control for variable speed variable pitch wind turbines. In *Proceedings of ACC* (pp. 2778–2784).
Spencer, M. D., Stol, K. A., Unsworth, C. P., Cater, J. E., & Norris, S. E. (2013). Model predictive control of a wind turbine using short-term wind field predictions. *Wind Energy*, *16*(3), 417–434.
Sutherland, H. J. (1999). *On the fatigue analysis of wind turbines* (Vol. 1999; Technical Report). Albuquerque, NM, and Livermore, CA: Sandia National Laboratories (SNL).
Taylor, G. I. (1938). The spectrum of turbulence. *Proceedings of the Royal Society A: Mathematical, Physical and Engineering Sciences*, *164*(919), 476–490.
Valencia-Palomo, G. (2010). *Efficient implementations of predictive control* (Ph.D. thesis). University of Sheffield.
van Engelen, T. G., & van der Hooft, E. L. (2005). *Individual pitch control inventory* 194 (Technical Report). ECN.
Vinnicombe, G. (2000). *Uncertainty and feedback*. London: Imperial College Press.
Wang, N., Johnson, K. E., Wright, A. D., & Wright, A. D. (2012). FX-RLS-based feedforward control for LIDAR-enabled wind turbine load mitigation. *IEEE Transactions on Control Systems Technology*, *20*(5), 1212–1222.

Chapter 3
Review of the Related Work

3.1 Overview

The aim of this chapter is to present the relevant literature regarding the blade-pitch control for wind turbine load reductions. The aim of this chapter is threefold. Firstly, the background and recent development of individual pitch control strategies are reviewed in Sect. 3.2. Secondly, in Sect. 3.3, an overview of the feed-forward model predictive control is presented and particularly, the shortcomings of the feed-forward compensator in standard MPC is highlighted. Lastly, Sect. 3.4 discusses the literature regarding the feed-forward control in wind turbines and followed by a chapter summary is concluded in Sect. 3.5. This chapter helps the reader to understand the recent developments and their shortcomings in literature, and most importantly, laying the foundation for the coming chapters.

3.2 Individual Pitch Control

As discussed in Chap. 1, a growing body of research has emerged in recent years, seeking to establish the best way of designing individual pitch control (IPC) systems. The key reason is the ability to actively control the pitch of each blade that offers the potential to reduce the unsteady loads that arise form a number of sources, such as wind-shear, tower shadow, yaw misalignment and turbulence within the atmospheric boundary layer (Barlas & van Kuik, 2010). Of the many IPC strategies that have been published in recent years, most can be grouped into two distinct classes, characterised by the specific turbine loads they are primarily designed to attenuate.

The first and most popular branch of IPCs is the *Coleman transform-based control*, that targets load reductions on the non-rotating turbine structures, such as the nacelle and main bearing. The second branch of IPCs is the *single-blade control* and the

W. H. (Alan) Lio, *Blade-Pitch Control for Wind Turbine Load Reductions*,
Springer Theses, https://doi.org/10.1007/978-3-319-75532-8_3

Clarke transform-based control, that targets load reductions upon the rotating turbine structures, primarily the blades. This section presents a thorough literature review regarding these three IPC design techniques.

3.2.1 Coleman Transform-Based Control

Coleman transform-based control is the earliest and a popular form of IPC strategy The Coleman transform-based IPC is depicted in Fig. 3.1, that is a part of the wind turbine architecture depicted in Fig. 2.16. The Coleman Transform $T_{\text{cm}}(\phi(t))$ is a time varying matrix that projects the rotational blade loads, $\tilde{M}_{1,2,3}$, onto the stationary and orthogonal tilt and yaw axes of the turbine, according to the blade azimuth angle $\phi(t)$. For a three-bladed turbine in which $\phi(t) = \omega(t)t$ is defined as the angle of the first blade from the horizontal yaw axis, the Coleman Transform is defined as follows:

$$\begin{bmatrix} \tilde{M}_{\text{tilt}}(t) \\ \tilde{M}_{\text{yaw}}(t) \end{bmatrix} := \underbrace{\frac{2}{3} \begin{bmatrix} \sin\phi(t) & \sin\left(\phi(t) + \frac{2\pi}{3}\right) & \sin\left(\phi(t) + \frac{4\pi}{3}\right) \\ \cos\phi(t) & \cos\left(\phi(t) + \frac{2\pi}{3}\right) & \cos\left(\phi(t) + \frac{4\pi}{3}\right) \end{bmatrix}}_{T_{\text{cm}}(\phi(t))} \begin{bmatrix} \tilde{M}_1(t) \\ \tilde{M}_2(t) \\ \tilde{M}_3(t) \end{bmatrix}, \tag{3.1a}$$

The tilt and yaw referred flap-wise blade root bending moments, $\tilde{M}_{\text{tilt}}$ and $\tilde{M}_{\text{yaw}}$ are mapped via the Coleman controller $C_{\text{cm}} \in \mathcal{R}^{2\times 2}$ to tilt and yaw referred pitch signals $\tilde{\theta}_{\text{tilt}}$ and $\tilde{\theta}_{\text{yaw}}$, that in turn are projected back into the blade referred pitch signals via the inverse Coleman Transform, $T_{\text{cm}}^{\text{inv}}(\phi(t))$ accordingly:

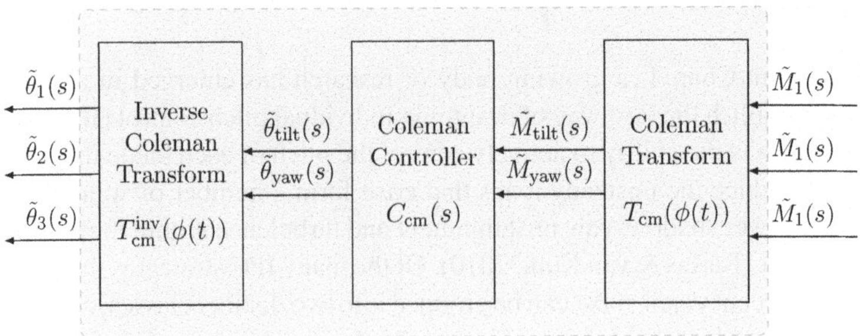

Fig. 3.1 Coleman Transform-based controller

$$\begin{bmatrix} \tilde{\theta}_1(t) \\ \tilde{\theta}_2(t) \\ \tilde{\theta}_3(t) \end{bmatrix} := \underbrace{\begin{bmatrix} \sin\phi(t) & \cos\phi(t) \\ \sin\left(\phi(t)+\frac{2\pi}{3}\right) & \cos\left(\phi(t)+\frac{2\pi}{3}\right) \\ \sin\left(\phi(t)+\frac{4\pi}{3}\right) & \cos\left(\phi(t)+\frac{4\pi}{3}\right) \end{bmatrix}}_{T_{\mathrm{cm}}^{\mathrm{inv}}(\phi(t))} \begin{bmatrix} \tilde{\theta}_{\mathrm{tilt}}(t) \\ \tilde{\theta}_{\mathrm{yaw}}(t) \end{bmatrix}, \tag{3.1b}$$

where $\tilde{\theta}_{1,2,3}$ are the perturbations in the blade pitch angle, as defined in Sect. 2.8 in Chap. 2.

These Coleman transformations $T_{\mathrm{cm}}(\phi(t))$ and $T_{\mathrm{cm}}^{\mathrm{inv}}(\phi(t))$ emerged from the area of helicopter rotor control (Coleman & Feingold, 1957) and also are widely employed in the fields of power conversion and electrical machines under the guise of the direct-quadrature-zero (dq0) or Park transformations (Park, 1929). Use of the Coleman transform to address the IPC problem in wind turbines was adopted by Caselitz et al. (1997) and Bossanyi (2003).

In the literature, there are many design methods for the Coleman controller $C_{\mathrm{cm}} \in \mathcal{R}^{2\times 2}$. Unlike the blade model (2.12) employed in this thesis, many Coleman controller design methods are relied upon different blade models. And most of these blade model can be grouped into three classes, that are as follows:

- Blade model from linearisation tools,
- Blade model with a diagonal structure,
- Blade model including the dynamics of the Coleman transformations.

3.2.1.1 Blade Model from Linearisation Tools

Instead of deriving the blade model from first principles like the linear differential equations (2.12) in Chap. 2, many studies use system identification techniques or linearisation codes to obtain a linear wind turbine model from a high-fidelity simulation turbine model, for example, BLADED (Bossanyi, 2003), TURBU (van Engelen & Braam, 2004) and FAST (Jonkman & Buhl Jr, 2005). For linear control design, these turbine models are linearised at a specific operating wind speed and a specific azimuth location of the blades (Jonkman & Buhl Jr, 2005) and this type of linear blade model is favoured by many studies because it is convenient and also the couplings between dynamics are modelled implicitly. For example, the earliest IPC study by Bossanyi (2003) employed a linear model generated from the BLADED code in order to design a linear quadratic Gaussian (LQG) regulator. Another example is Stol, Zhao, and Wright (2006), where a gain-scheduling linear quadratic regulator (LQR) controller was constructed based on an azimuth-dependent blade model. Nonetheless, the linear model generated from linearisation codes might be too complicated, so the turbine dynamics and structural couplings behind these linear models cannot be easily understood, making linear analysis difficult. Thus, studies started a 'bottom-up' approach by modelling the blade using first principles (e.g. van Engelen & van der Hooft, 2005).

3.2.1.2 Blade Model with a Diagonal Structure

Some studies (e.g. Selvam, 2007; Selvam, Kanev, van Wingerden, van Engelen, & Verhaegen, 2009; van Engelen & van der Hooft, 2005) derived a linear wind turbine model including the rotor, tower and blade dynamics from first principles. Notice that in this turbine model, the blade is assumed to be rigid. Thus, unlike the dynamic blade model (2.12), the blade dynamics is described by a simple static relationship, mapping the pitch angle to the blade moment. For example, in van Engelen and van der Hooft (2005) and Selvam (2007), the linear blade model is of the following form:

$$\tilde{M}_i(t) = \frac{dM_{\text{flap}}}{d\theta}\tilde{\theta}_i(t), \tag{3.2}$$

where $\frac{dM_{\text{flap}}}{d\theta} \in \mathbb{R}$ are the variations of the blade moments to the pitch angle. Subsequently, substituting the Coleman transformations (3.1) into the rigid blade model (3.2) yields (e.g. Selvam, 2007; van Engelen & van der Hooft, 2005):

$$\tilde{M}_{\text{tilt}}(t) = \frac{dM_{\text{flap}}}{d\theta}\tilde{\theta}_{\text{tilt}}(t), \quad \tilde{M}_{\text{yaw}}(t) = \frac{dM_{\text{flap}}}{d\theta}\tilde{\theta}_{\text{yaw}}(t). \tag{3.3}$$

Thus, the Coleman-transformed plant is of a *diagonal structure* as follows:

$$\begin{bmatrix} \tilde{M}_{\text{tilt}}(s) \\ \tilde{M}_{\text{yaw}}(s) \end{bmatrix} = \begin{bmatrix} G_{\text{tilt}}(s) & 0 \\ 0 & G_{\text{yaw}}(s) \end{bmatrix} \begin{bmatrix} \tilde{\theta}_{\text{tilt}}(s) \\ \tilde{\theta}_{\text{yaw}}(s) \end{bmatrix}. \tag{3.4}$$

For illustration purpose, the tower and rotor dynamics are not included in (3.4). However, this does not affect the decoupled structure of the blade model. This form of the Coleman-transformed blade model (3.4) is favoured by many studies. For example, van Engelen and van der Hooft (2005) designed a diagonal form of the tilt and yaw controller for attenuating the blade referred loads upon the tilt and yaw turbine axes. The tilt and yaw controller is of the following diagonal form:

$$\begin{bmatrix} \tilde{\theta}_{\text{tilt}}(s) \\ \tilde{\theta}_{\text{yaw}}(s) \end{bmatrix} = \begin{bmatrix} K_{\text{tilt}}(s) & 0 \\ 0 & K_{\text{yaw}}(s) \end{bmatrix} \begin{bmatrix} \tilde{M}_{\text{tilt}}(s) \\ \tilde{M}_{\text{yaw}}(s) \end{bmatrix}. \tag{3.5}$$

where the controllers $K_{\text{tilt}}, K_{\text{yaw}} \in \mathcal{R}$ are typically a PID design. The *diagonal controller* design (3.5) is employed by many studies. For example, Bossanyi (2005), Jelavic, Petrovic, and Peric (2008), Lackner and van Kuik (2010) and Plumley, Leithead, Jamieson, Bossanyi, and Graham (2014).

Nonetheless, the Coleman transformations (3.1) only projects the blade loads at 1p (once-per-revolution) frequency into the turbine tilt and yaw axes as demonstrated by van Engelen (2006). If the Coleman transformations (3.1) is modulated with multiples of the rotational frequencies, for example, replacing $\omega(t)$ in (3.1) with $2\omega(t)$, the blade loads at higher harmonics can be also mapped as static loads on the turbine tilt and yaw axes. Subsequently, these referred static loads upon the tilt and yaw axes can be attenuated by the diagonal controller (3.5). Similar methods were also adopted by Bottasso, Croce, Riboldi, and Nam (2013).

Furthermore, to improve the robustness to the model uncertainties, for example, couplings between the tower and blade structures (Leithead, Dominguez, & Spruce, 2004), a multi-variable $\mathcal{H}_\infty$ design was proposed by Geyler and Caselitz (2007, 2008).

Comparing the code-generated linear blade model, the major difference is that the couplings between the tilt and yaw axes are not modelled in (3.4), as shown by studies (Selvam, 2007; Selvam et al., 2009). Later, this inspired work by Lu, Bowyer, and Jones (2015), revealing that despite the diagonal controller (3.5) based on the decoupled form of the blade model (3.4) yielding a good gain and phase margin, but lacking the dynamic couplings between the tilt and yaw loop, the diagonal controller (3.5) could perform poorly, even resulting in closed-loop instability, upon the *blade model including the dynamics of the Coleman transformations*. Thus, Lu et al. (2015) investigated the influence of the Coleman transformations on the system dynamics.

3.2.1.3 Blade Model Including the Dynamics of the Coleman Transformations

Many of the Coleman transform-based control studies modelled the blade model as (3.2), where the tilt and yaw loop are decoupled. A recent study by Lu et al. (2015) demonstrated that a decoupled tilt and yaw control design, such as (3.5), could be poor if the couplings between the tilt and yaw loop were not considered in the modelling explicitly. Thus, to investigate the influence of the Coleman transformations on the blade dynamics, Lu et al. (2015) models the blade as a second-order system and taking into account the actuator and filter dynamics. Consider the dynamics of the blade as follows:

$$\tilde{M}_i(s) = G(s)\tilde{\theta}_i(s) \tag{3.6}$$

where the transfer function $G \in \mathcal{R}$ is the same as (2.52) in Chap. 2. The Coleman transformations (3.1) modify the plant (3.6) to yield the Coleman-transformed model as follows:

$$\begin{bmatrix} \tilde{M}_{\text{tilt}}(s) \\ \tilde{M}_{\text{yaw}}(s) \end{bmatrix} = \begin{bmatrix} \dfrac{G(s+j\omega_0)+G(s-j\omega_0)}{2} & j\dfrac{G(s+j\omega_0)-G(s-j\omega_0)}{2} \\ -j\dfrac{G(s+j\omega_0)-G(s-j\omega_0)}{2} & \dfrac{G(s+j\omega_0)+G(s-j\omega_0)}{2} \end{bmatrix} \begin{bmatrix} \tilde{\theta}_{\text{tilt}}(s) \\ \tilde{\theta}_{\text{yaw}}(s) \end{bmatrix}, \tag{3.7}$$

where $\omega_0 \in \mathbb{R}$ is the constant rated rotor speed, and from which the coupled nature of the tilt and yaw loops is evident. The derivation of (3.7) is similar to the proof of Lemma 4.1 in Chap. 4, thus, the details are not included here. Lu et al. (2015) subsequently designed a $\mathcal{H}_\infty$ loop-shaping controller, based on (3.7), that outperformed a comparative diagonal controller (3.5). This form of the linear Coleman-transformed plant (3.7) was later adopted by another recent study by Vali, van Wingerden, and Kuhn (2016) for a mixed sensitivity $\mathcal{H}_\infty$ optimisation approach.

Remark 3.1 This thesis adopts the blade model (3.6) and Coleman-transformed plant (3.7) by Lu et al. (2015) because these models take into account the frequency-shifting effect of the Coleman transformations, which is particularly useful when analysing the similarities between the Coleman transform-based control with other types of IPCs.

Apart from the theoretical development, the Coleman transform-based control strategies have also been field-tested on a number of wind turbines. As reported by Bossanyi and Wright (2009) and Bossanyi et al. (2012), the IPC controllers by Bossanyi (2003) have been implemented on a 660 kW two-bladed wind turbine. A similar field test was also conducted on a 600 kW three-bladed wind turbine (Bossanyi, Fleming, & Wright, 2013). In recent years, an IPC algorithm by Jasniewicz and Geyler (2011) was also tested upon a 5MW wind turbine (Shan & Adelt, 2013).

3.2.2 *Single-Blade Control*

One of the simplest forms of IPC strategies is the single-blade control and this concept was originated by Leithead, Neilson, Dominguez, and Dutka (2009) and Leithead, Neilson, and Dominguez (2009). As depicted in Fig. 3.2, the single-blade control, also later termed as individual blade control (Han & Leithead, 2014), equips each blade with its own controller for attenuating the blade, where the blade controller actuates in response to its local blade measurements. Unlike the Coleman transform-based control, the single-blade control does not involve any coordinate transformations, and this type of IPC strategies has a decoupled structure, consisting of three SISO controllers, as follows:

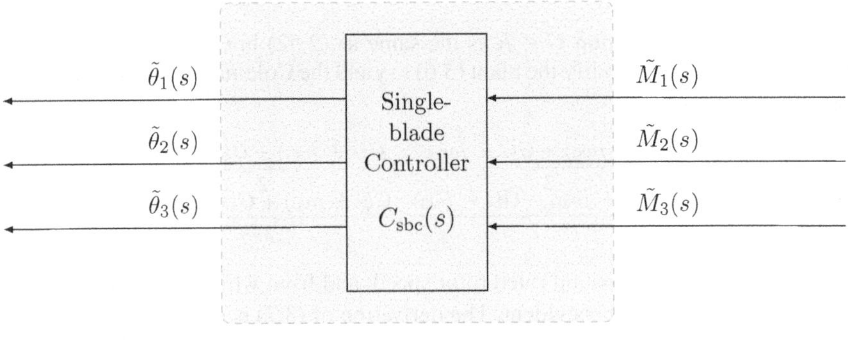

Fig. 3.2 Single-blade controller

$$\begin{bmatrix} \tilde{\theta}_1(s) \\ \tilde{\theta}_2(s) \\ \tilde{\theta}_3(s) \end{bmatrix} = \underbrace{\begin{bmatrix} K(s) & 0 & 0 \\ 0 & K(s) & 0 \\ 0 & 0 & K(s) \end{bmatrix}}_{C_{\text{sbc}}(K(s))} \begin{bmatrix} \tilde{M}_1(s) \\ \tilde{M}_2(s) \\ \tilde{M}_3(s) \end{bmatrix} \quad (3.8)$$

where the blade controller $K \in \mathcal{R}$ is designed to attenuate the blade loads at 1p, 2p and 4p frequencies. This form of IPC (3.8) is investigated by many studies, for example, Yi and Leithead (2012) and Han and Leithead (2014) examined the performance of the single-blade controller in terms of fatigue and ultimate load reductions.

Remark 3.2 In this thesis, the baseline IPC is a form of the single-blade controller (3.8) for its convenience as no coordination is required.

3.2.3 Clarke Transform-Based Control

A typical Clarke transform-based control is depicted in Fig. 3.3. The Clarke transformations, also known as $\alpha\beta\gamma$ transformations, are widely used in the fields of electrical machines (Duesterhoeft, Schulz, & Clarke, 1951). These transformations were adopted by Zhang, Cheng, and Chen (2013) in the IPC design for wind turbines. In the Clarke transform-based control, the blade moment measurements $\tilde{M}_{1,2,3}$ are mapped into a set of orthogonal axes that are stationary to the rotating turbine blades via the Clarke transform $T_{\text{ck}} \in \mathbb{R}^{2\times3}$, that is defined as follows:

$$\begin{bmatrix} \tilde{M}_\alpha(s) \\ \tilde{M}_\beta(s) \end{bmatrix} = \underbrace{\sqrt{\frac{2}{3}} \begin{bmatrix} 1 & -\frac{1}{2} & -\frac{1}{2} \\ 0 & \frac{\sqrt{3}}{2} & -\frac{\sqrt{3}}{2} \end{bmatrix}}_{T_{\text{ck}}} \begin{bmatrix} \tilde{M}_1(s) \\ \tilde{M}_2(s) \\ \tilde{M}_3(s) \end{bmatrix}, \quad (3.9a)$$

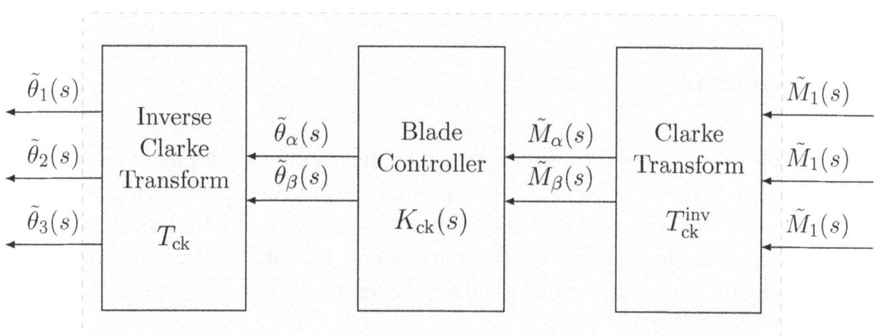

Fig. 3.3 Clarke Transform-based controller

where $\tilde{M}_\alpha$, $\tilde{M}_\beta$ denote the referred blade signals upon the orthogonal axes. Similarly, the referred pitch signals upon the orthogonal axes $\tilde{\theta}_\alpha$, $\tilde{\theta}_\beta$ are projected back to the rotating frame of reference via the inverse Clarke transform $T_{\mathrm{ck}}^{\mathrm{inv}} \in \mathbb{R}^{3\times 2}$, that is defined as follows:

$$\begin{bmatrix} \tilde{\theta}_\alpha(s) \\ \tilde{\theta}_\beta(s) \end{bmatrix} = \underbrace{\sqrt{\frac{2}{3}} \begin{bmatrix} 1 & 0 \\ -\frac{1}{2} & \frac{\sqrt{3}}{2} \\ -\frac{1}{2} & -\frac{\sqrt{3}}{2} \end{bmatrix}}_{T_{\mathrm{ck}}^{\mathrm{inv}}} \begin{bmatrix} \tilde{\theta}_1(s) \\ \tilde{\theta}_2(s) \\ \tilde{\theta}_3(s) \end{bmatrix}. \tag{3.9b}$$

As demonstrated by Zhang et al. (2013), substituting the Clarke transformations (3.9) into the blade model (3.6) yields:

$$\begin{bmatrix} \tilde{M}_\alpha(s) \\ \tilde{M}_\beta(s) \end{bmatrix} = \begin{bmatrix} G(s) & 0 \\ 0 & G(s) \end{bmatrix} \begin{bmatrix} \tilde{\theta}_\alpha(s) \\ \tilde{\theta}_\beta(s) \end{bmatrix}. \tag{3.10}$$

Based on the blade model (3.10), which has a diagonal structure, a diagonal blade controller $K_{\mathrm{ck}} \in \mathcal{R}^{2\times 2}$ can be employed, that is defined as follows:

$$\begin{bmatrix} \tilde{\theta}_\alpha(s) \\ \tilde{\theta}_\beta(s) \end{bmatrix} = \underbrace{\begin{bmatrix} K(s) & 0 \\ 0 & K(s) \end{bmatrix}}_{K_{\mathrm{ck}}(K(s))} \begin{bmatrix} \tilde{M}_\alpha(s) \\ \tilde{M}_\beta(s) \end{bmatrix}. \tag{3.11}$$

and the controller $K \in \mathcal{R}$ targets the blade loads at 1p, 2p and 4p frequencies. Notice that the use of the Clarke transformations simplifies the three single-input-single-output (SISO) controller structure (3.8), from the single-blade control, into two SISO control structure (3.11). Furthermore, unlike the Coleman transform-based IPC, the Clarke transform-based control does not require a measurement of the rotor azimuth angle, where the same benefit also holds for single-blade control.

3.2.4 Summary

This section presented background regarding the recent development of the IPC strategies. The features of each IPC design method are discussed, laying the foundation for performance similarity analysis in Chap. 4. In addition, this section also explained the influence of the Coleman transformation on the blade model, that justifies why the Coleman-transformed plant (3.7) is employed to represent the Coleman transform-based control in Chap. 4.

3.3 Feed-Forword Model Predictive Control

As discussed in Chap. 1, one of the objectives in this thesis is to incorporate the upcoming wind information and constraint handling into the existing blade-pitch controller. The aim of this section is to present the development of the feed-forward MPC that particularly handles advance information. Notice that the standard MPC performs poorly in terms of handling the advance knowledge, and many studies suggested a separate feed-forward design upon a feedback design for improving the tracking or disturbance performance. Methods of incorporating feed-forward into the feedback MPC design can be grouped into two classes: the *generalised predictive control (GPC)* and *dual-mode MPC*.

3.3.1 Feed-Forward in Generalised Predictive Control

In Chap. 2, the background of generalised predictive control (GPC) was presented. The reader may expect, in principle, the standard GPC (Clarke, Mohtadi, & Tu, 1987a, 1987b), should automatically take into account the feed-forward information $\underset{\rightarrow k}{r}$ within the optimisation (2.24), resulting in an unconstrained control law (2.21):

$$\Delta u_k = -Kx_k + P_r \underset{\rightarrow k}{r}. \tag{3.12}$$

However, the performance of the feed-forward compensator P_r could be poor (Rossiter & Grinnell, 1995, 1996), explained as follows.

Studies by Rossiter and Grinnell (1995, 1996) argued that in practice, the industry tends to deploy a small control horizon n_c in the concern of the computational power. As a result, the performance of the default feed-forward compensator P_r in (3.12) is poor. To illustrate, consider a simple step-tracking example,

$$x_{k+1} = \underbrace{\begin{bmatrix} 0.8 & 0.2 \\ -0.2 & 0.9 \end{bmatrix}}_{A} x_k + \underbrace{\begin{bmatrix} 0.1 \\ 0.8 \end{bmatrix}}_{B} u_k, \; y_k = \underbrace{\begin{bmatrix} 1.9 & -1 \end{bmatrix}}_{C} x_k. \tag{3.13}$$

Figure 3.4 shows the performance of the controller law (3.12) with the control horizon $n_c = 1, 10$ whilst the prediction and preview horizons are fixed at $n = 10, n_a = 10$. The set-point changes happen at sample $k = 20$. For the case with the small control horizon $n_c = 1$, the control move u_k reacted too soon before the set-point r_k changes occur, whilst for the controller with the control horizon $n_c = 10$, the performance is much better. Thus, it is clearly the default feed-forward compensator P_r in (3.12) is not well designed when the GPC is tuned with small n_c.

Based on the findings from Rossiter and Grinnell (1995, 1996), many studies began investigating design methods of improving the feed-forward compensator P_r, separating from the GPC feedback design (e.g. Rossiter & Valencia-Palomo, 2009;

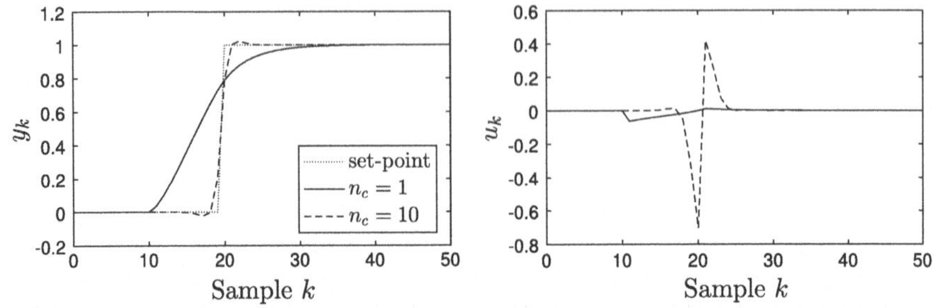

(A) Output Response. (B) Input Response.

Fig. 3.4 Response of the GPC with various control horizon n_c. The prediction and preview horizon, n and n_a, both are set as 10. Solid and dashed lines represent the GPC with $n_c = 1$ and $n_c = 10$, respectively, whilst the dotted line shows the set-point

Valencia-Palomo, Rossiter, & López-Estrada, 2014; Wang & Rossiter, 2008). For example, in Rossiter and Valencia-Palomo (2009), a separate *two stage design* was proposed. Consider an unconstrained control law (2.21):

$$\Delta u_k = -Kx_k + P_r \underset{\rightarrow k}{r}. \tag{3.14}$$

In Rossiter and Valencia-Palomo (2009), the default feed-forward compensator P_r is optimised based on the *closed-loop predictions* of the input and output, that can be obtained by substituting (3.14) into the prediction model (2.19):

$$\underbrace{\begin{bmatrix} y_{1|k} \\ y_{2|k} \\ \vdots \\ y_{n|k} \end{bmatrix}}_{\underset{\rightarrow k}{y}} = \begin{bmatrix} C\Phi \\ C\Phi^2 \\ \vdots \\ C\Phi^n \end{bmatrix} x_{0|k} + \underbrace{\begin{bmatrix} CB & 0 & \cdots & 0 \\ C\Phi B & CB & \cdots & 0 \\ \vdots & \vdots & \ddots & \vdots \\ C\Phi^{n-1}B & C\Phi^{n-2}B & \cdots & C\Phi^{n-n_c}B \end{bmatrix}}_{H_y} \underbrace{\begin{bmatrix} \underset{\rightarrow k-1}{r}^T \\ \underset{\rightarrow k}{r}^T \\ \vdots \\ \underset{\rightarrow k+n_c-2}{r}^T \end{bmatrix}}_{S_r} \underbrace{\begin{bmatrix} P_1 \\ P_2 \\ \vdots \\ P_{n_a} \end{bmatrix}}_{P_r}, \tag{3.15a}$$

$$\underbrace{\begin{bmatrix} \Delta u_{0|k} \\ \Delta u_{1|k} \\ \vdots \\ \Delta u_{n_c-1|k} \end{bmatrix}}_{\Delta \underset{\rightarrow k-1}{u}} = -\begin{bmatrix} K \\ K\Phi \\ \vdots \\ K\Phi^{n_c-2} \end{bmatrix} x_{0|k} + \underbrace{\begin{bmatrix} I & 0 & \cdots & 0 \\ B & I & \cdots & 0 \\ \vdots & \vdots & \ddots & \vdots \\ \Phi^{n_c-3}B & \Phi^{n_c-4}B & \cdots & I \end{bmatrix}}_{H_u} \underbrace{\begin{bmatrix} \underset{\rightarrow k-1}{r}^T \\ \underset{\rightarrow k}{r}^T \\ \vdots \\ \underset{\rightarrow k+n_c-2}{r}^T \end{bmatrix}}_{S_r} \underbrace{\begin{bmatrix} P_1 \\ P_2 \\ \vdots \\ P_{n_a} \end{bmatrix}}_{P_r}, \tag{3.15b}$$

where $\Phi = A - BK$. Assuming zero initial condition $x_{0|k} = 0$, the default feed-forward compensator P_r is optimised by minimising the cost (2.20a) with respect to the P_r, yielding:

$$P_r = \left(S_r^T H_y^T Q H_y S_r + S_r^T H_u^T Q H_u S_r\right)^{-1} S_r^T H_y^T \underset{\rightarrow k-1}{r}, \tag{3.15c}$$

Finally, the new feed-forward compensator P_r in (3.15c) replaces the default compensator in (3.14). Notice that P_r in (3.15c) is dependent on the upcoming set-point $\underset{\rightarrow k-1}{r}$ at sample k. Thus, computation of the P_r in (3.15c) requires solving a least-square problem on-line at every sample. In Valencia-Palomo et al. (2014), the study demonstrated the performance of this two stage design method on hardware that showed a significant improvement of using the re-designed feed-forward compensator.

Similar studies also investigated the separate design under the GPC framework. Instead of improving the default feed-forward compensator P_r on an existing feedback GPC design, (Carrasco & Goodwin, 2011a, 2011b; Goodwin, Carrasco, & Mayne, 2011) argued that assuming the feed-forward control law is not limited by the feedback bandwidth, thus, the feed-forward controller should take an active role to perform the tracking or disturbance rejection without any measurement from the plant, whilst the feedback design is constructed for robustifying the closed-loop. To illustrate, consider the cost function for the feed-forward control law:

$$\arg \min_{\underset{\rightarrow k-1}{\Delta u^{\text{ff}}}} J_k^{\text{ff}} = \sum_{i=1}^{n} \left((r_{i|k} - y_{i|k}^{\text{ff}})^T Q (r_{i|k} - y_{i|k}^{\text{ff}}) + (\Delta u_{i-1|k}^{\text{ff}})^T R \Delta u_{i-1|k}^{\text{ff}} \right), \tag{3.16a}$$

$$\text{s.t.}\quad x_{i+1|k}^{\text{ff}} = A x_{i|k}^{\text{ff}} + B \Delta u_{i-1|k}^{\text{ff}},\ \forall i = \{1, 2, \ldots, n\}, \tag{3.16b}$$

$$y_{i|k}^{\text{ff}} = C x_{i|k}^{\text{ff}},\ \forall i = \{1, 2, \ldots, n\}. \tag{3.16c}$$

Notice that the optimisation problem of (3.16) needs to take into account constraints such as (2.23). It is assumed that the initial condition $x_{0|k}^{\text{ff}}$ is known. Once the feed-forward control input sequence $\underset{\rightarrow k-1}{\Delta u^{\text{ff}}}$ is computed, the entire sequence is then passed into the feedback optimisation:

$$\arg \min_{\underset{\rightarrow k-1}{\Delta u^{\text{fb}}}} J_k^{\text{fb}} = \sum_{i=1}^{n} \left((r_{i|k} - y_{i|k})^T Q (r_{i|k} - y_{i|k}) + (\Delta u_{i-1|k}^{\text{fb}})^T R \Delta u_{i-1|k}^{\text{fb}} \right), \tag{3.17a}$$

$$\text{s.t.}\quad x_{i|k} = Ax_{i|k} + B(\Delta u^{\text{fb}}_{i-1|k} + \Delta u^{\text{ff}}_{i-1|k}),\ \forall i = \{1, 2, \ldots, n\}, \tag{3.17b}$$

$$y_{i|k} = Cx_{i|k},\ \forall i = \{1, 2, \ldots, n\}, \tag{3.17c}$$

and the optimisation (3.17) also is subjected to constraints like (2.23). Notice that this design requires solving two quadratic programming (QP) problems on-line. Unlike the two stage design in (3.15c), the approach in (3.16) and (3.17) is similar to a model inversion-based approach and if the limitation on the bandwidth is mainly due to the input hardware constraints, this approach has the same performance as if the feed-forward optimisation is not included as pointed out by the authors in (Carrasco & Goodwin, 2011a).

3.3.2 Feed-Forward in Dual-Mode MPC

In dual-mode MPC, the use of an infinite-horizon cost function makes formulating the feed-forward knowledge difficult. The standard dual-mode MPC relies upon an assumption that the set-point is fixed beyond the control horizon n_c (e.g. Kouvaritakis, Rossiter, & Cannon, 1998; Scokaert & Rawlings, 1998). By inspection of the dual-mode cost in (2.38):

$$J_k := \underset{\rightarrow k}{\tilde{x}}^T \bar{Q}_x \underset{\rightarrow k}{\tilde{x}} + \Delta \underset{\rightarrow k-1}{\tilde{u}}^T \bar{R} \Delta \underset{\rightarrow k-1}{\tilde{u}}, \tag{3.18}$$

where the prediction of state $\underset{\rightarrow k}{\tilde{x}}$ and input $\Delta \underset{\rightarrow k-1}{\tilde{u}}$ both are up to the control horizon n_c. Thus, undoubtedly, the changes in the set-point beyond the control horizon n_c can not be handled properly by the cost formulation in the standard dual-mode MPC (3.18).

Nonetheless, the reader may wonder that the steady-state x^s_k, Δu^s_k in (2.29) takes into account the knowledge of the set-point r_k. Consider the unconstrained control law (2.27):

$$\Delta u_k = -K(x_k - x^s_k) + \Delta u^s_k, \tag{3.19}$$

where $x^s_k := K_{xr} r_k$ and $\Delta u^s_k := K_{ur} r_k$. As argued by Rossiter and Valencia-Palomo (2009), despite the set-point knowledge is incorporated into the steady-state x^s_k, Δu^s_k, however, the set-point trajectory $\underset{\rightarrow k-1}{r}$ are not fully utilised. This is because the gain of the steady-state K_{xr}, K_{ur} only ensure the zero offset and any changes in the set-point trajectory are not handled systematically.

This issue becomes clearer in the closed-loop paradigm. Consider the cost function in the closed-loop paradigm from (2.41):

$$J_k = \underset{\rightarrow k-1}{c}^T \bar{W}_c \underset{\rightarrow k-1}{c}. \tag{3.20}$$

In absence of constraints, the control law is simply (3.19) with $c_k = 0$ even if the set-point trajectory is not constant. Hence, no input efforts are made for the changes happened in the set-point trajectory $\underset{\rightarrow k-1}{r}$.

To overcome this issue, the two stage design (3.15), as discussed in Sect. 3.3.1, can be extended to the dual-mode paradigm (Rossiter & Valencia-Palomo, 2009; Valencia-Palomo et al., 2014), where a feed-forward compensator P_r can be computed based on the closed-loop predictions, which can be obtained by (3.15). Another method proposed by Rossiter (2015) was to embed the set-point changes $\underset{\rightarrow k-1}{r}$ into the cost function (3.20). This method has many benefits in that only one on-line optimisation is required compared to the two stage design in (3.15) and no assumptions are placed on the initial conditions, thus, it is adopted by this thesis to incorporate the upcoming wind information into the existing feedback controller, as discussed in Chap. 6.

Remark 3.3 In the literature, the feed-forward information is mainly considered as the set-point trajectory, whilst in this thesis, the advance knowledge is the upcoming wind measurement, which is an input disturbance to the turbine system. However, the feed-forward issues, discussed in this section, also apply to controllers, that are regardless if their control purposes are either tracking or disturbance rejection.

3.3.3 *Summary*

This section presented the problems of handling upcoming knowledge in the GPC and dual-mode MPC. Many studies demonstrated methods to improve the feed-forward design based on a feedback design under the MPC framework. However, most of the feedback design is considered as a state-feedback control law. In Chap. 6, a method is proposed to incorporate feed-forward knowledge and constraint handling capability into an output-feedback controller. In addition, the difficulties regarding such an approach are also discussed in Chap. 6.

3.4 Feed-Forward Control in Wind Turbines

As discussed in Chap. 1, load attenuation strategies via active blade pitching are effective in reducing unsteady loads on the blades and rotor. In recent years, many studies began utilising the upcoming wind measurements for preview blade pitch control design, with a view to compensating the wind disturbance before it strikes the turbine. This section presents some relevant works on the feed-forward control on wind turbine, grouping into three aspects: *modelling of the wind field*, *feed-forward design methods* and *wind measurement quality*.

3.4.1 Modelling of the Wind Field

Typically, wind field information can be represented in a collection of vectors in a Cartesian or polar grid (Schlipf, Schuler, Grau, & Martin, 2010). For example, the TurbSim, discussed in Sect. 2.8, generates a series of three dimensional wind vectors on a two-dimensional grid, marching towards the turbine. However, the complex wind field with multi-dimensional is not suitable for control design, because typical control design relies upon simpler models, for example, the linear model shown in Sect. 2.4. Thus, many methods were proposed to simplify the wind disturbance.

3.4.1.1 Rotor Speed Regulation

Early adoption of preview control in wind turbines for turbine speed regulation was reported by Kodama, Matsuzaka, Tuchiya, and Arinaga (1999), in which the preview control strategy was based on the *hub-height wind measurement* taken 40 m in front of the rotor by an anemometer on a free standing tower. In recent years, light detection and ranging (LIDAR) devices became commercially accessible and many studies (e.g. Harris, Hand, & Wright, 2005; Schlipf & Kuhn, 2008) began utilising the LIDAR devices for improving rotor speed regulation. Notice that there are many issues within the LIDAR devices, for example, how to improve the quality of the wind measurements (Simley, Pao, & Frehlich, 2011) and this is discussed later in Sect. 3.4.3. In this section, it is implicitly assumed that the wind information is available in a collection of upcoming stream-wise wind speed on a two-dimensional grid in front of the turbine.

Consider the measured stream-wise wind speed defined as $v_{\mathrm{m}}(y_{\mathrm{r}}(n), z_{\mathrm{r}}(n), n, t)$, where $y_{\mathrm{r}}(n), z_{\mathrm{r}}(n) \in \mathbb{R}$ denote the horizontal and vertical co-ordinates across the rotor plane and $n \in \mathbb{R}$ is the index of the wind data. To simplify the grid of the measured wind speed $v_{\mathrm{m}}(y_{\mathrm{r}}(n), z_{\mathrm{r}}(n), n, t)$ for control design, Schlipf (2013) suggested a simplest wind disturbance model, that is averaging all the available measurements, and defined as follows:

$$v_{\mathrm{eff}}(t) := \frac{1}{N} \sum_{n=1}^{N} v_{\mathrm{m}}(y_{\mathrm{r}}(n), z_{\mathrm{r}}(n), n, t), \tag{3.21}$$

where $N \in \mathbb{R}$ is the total number of measurement point. The rotor effective wind speed v_{eff} is then employed by the collective pitch controller for regulating the rotor speed. Nonetheless, given that the geometry of the blade is not uniform but irregular along the blade length, thus, the sensitivity to the wind along the blade is different. Simply averaging the measured wind speed (3.21) might not produce a good representation of the wind field.

Therefore, Schlipf and Kuhn (2008) and Schlipf, Fischer, and Carcangiu (2010) proposed a new method to calculate the rotor effective wind speed, where this effective speed is calculated based on the measured wind speeds across the rotor and the weightings of the variations of blade in-plane forces to wind speed. Consider the variations of blade in-plane forces to wind speed to wind speed defined as $\frac{F_{\text{in}}^{i,l}}{dv} \in \mathbb{R}$ and the steam-wise wind speed $v_{i,l} \in \mathbb{R}$ on the blade element $l \in \{1, \ldots, L\} \subset \mathbb{Z}$ along blade $i \in \{1, 2, 3\}$, the rotor effective wind speed v_{eff} is calculated as follows:

$$v_{\text{eff}}(t) := \frac{\sum_{i=1}^{3} \sum_{l=1}^{L} \left(\frac{F_{\text{in}}^{i,l}}{dv} v_{i,l}(t) r_l \right)}{\sum_{i=1}^{3} \sum_{l=1}^{L} \frac{F_{\text{in}}^{i,l}}{dv} r_l}, \tag{3.22}$$

where $r_l \in \mathbb{R}$ is the blade length. Notice that in Sect. 2.4, the wind-induced rotor disturbance $\tau_{\text{a}}^{\text{d}}$ (2.11) is similar to the rotor effective wind speed v_{eff} in (3.22) except where the rotor effective wind speed v_{eff} is divided by the term $\sum_{i=1}^{3} \sum_{l=1}^{L} \frac{F_{\text{in}}^{i,l}}{dv} r_l$.

3.4.1.2 Blade Load Attenuations

One of the earliest adoptions of the upcoming wind measurements for the blade load attenuation was by Laks, Pao, and Wright (2009). The study assumed the advance wind measurements are available in front of the blade tip, at 100% span of the blades. Later, this method is adopted by Dunne, Pao, Wright, Jonkman, and Kelley (2011), Laks, Pao, Wright, Kelley, and Jonkman (2010, 2011) considering the upcoming wind measurements at 75% span of the blades. However, the drawback of this method is that any additional wind field information across the rotor becomes redundant.

Therefore, Schlipf (2013), Schlipf, Schuler, et al. (2010) proposed a method that simplifies the wind field across the rotor into *three wind components*, the averaged wind speed $\bar{v} \in \mathbb{R}$, horizontal and vertical wind shear components $\delta_{\text{h}}, \delta_{\text{v}} \in \mathbb{R}$, respectively. Thus, consider the measured stream-wise wind speed $v_{\text{m}}(y_{\text{r}}(n), z_{\text{r}}(n), n, t)$, the simplified stream-wise wind speed $v(y_{\text{r}}(n), z_{\text{r}}(n), n, t)$ is defined as follows:

$$v(y_{\text{r}}(n), z_{\text{r}}(n), n, t) := \bar{v}(t) + y_{\text{r}}(n)\delta_{\text{h}}(t) + z_{\text{r}}(n)\delta_{\text{v}}(t), \tag{3.23}$$

where $y_{\text{r}}(n), z_{\text{r}}(n) \in \mathbb{R}$ denote the horizontal and vertical co-ordinates across the rotor plane and $n \in \{1, \ldots, N\} \subset \mathbb{Z}$ is the index of the wind data.The averaged wind speed $\bar{v}$, horizontal and vertical wind shear components $\delta_{\text{h}}, \delta_{\text{v}}$ can be computed by solving the following equation:

$$\begin{bmatrix} 1 & y_{\text{r}}(1) & z_{\text{r}}(1) \\ 1 & y_{\text{r}}(2) & z_{\text{r}}(2) \\ \vdots & \vdots & \vdots \\ 1 & y_{\text{r}}(N) & z_{\text{r}}(N) \end{bmatrix} \begin{bmatrix} \bar{v}(t) \\ \delta_{\text{h}}(t) \\ \delta_{\text{v}}(t) \end{bmatrix} = \begin{bmatrix} v_{\text{m}}(y_{\text{r}}(1), z_{\text{r}}(1), 1, t) \\ v_{\text{m}}(y_{\text{r}}(2), z_{\text{r}}(2), 2, t) \\ \vdots \\ v_{\text{m}}(y_{\text{r}}(N), z_{\text{r}}(N), N, t) \end{bmatrix}. \tag{3.24}$$

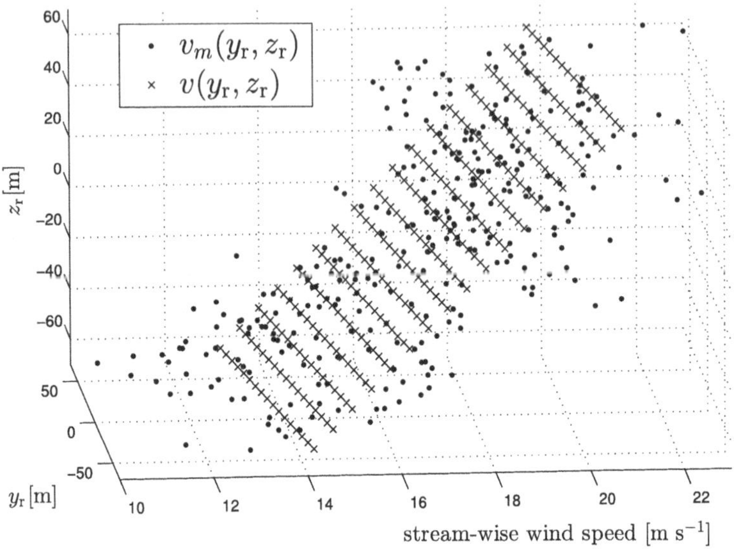

Fig. 3.5 Simulated wind field across the rotor plane. The measured stream-wise wind speed is denoted as v_m, whilst v represents the simplified wind field

An unique solution for $\bar{v}$, δ_h, δ_v can be found if $N \geq 3$ by simple least square method. Figure 3.5 demonstrates the difference between the simplified stream-wise wind speed v and the measured wind speed v_m. The simplified wind model (3.23) is popular and employed by many studies (e.g. Laks et al., 2011; Spencer, Stol, Unsworth, Cater, & Norris, 2013), because there is a linear blade model generated by the FAST linearisation code (Jonkman & Buhl Jr, 2005), where the three wind component representations of the wind field can be used as the input. Nonetheless, a comparison study showed that the load turbine reduction performance based on the simplified wind model (3.23) was worse than the direct wind measurements at 75% span of the blades (Laks et al., 2010, 2011). This is because the details of the turbulence of the wind field is averaged out by the three wind component representation (3.23).

Later, Simley, Dunne, Laks, and Pao (2013), Simley and Pao (2013) introduced the concept of the *blade effective wind speed*, which is similar to the rotor effective wind speed as discussed in the previous section. The blade effective wind speed v_{eff_i} is a weighted sum of the wind speed along the blade, which is defined as follows:

$$v_{\text{eff}_i}(t) := \frac{\sum_{l=1}^{L}\left(\frac{F_{\text{out}}^{i,l}}{dv} v_{i,l}(t) r_l\right)}{\sum_{l=1}^{L} \frac{F_{\text{in}}^{i,l}}{dv} r_l}, \tag{3.25}$$

where $v_{i,l}$ is the measured stream-wise wind speed, whilst $\frac{F_{\text{out}}^{i,l}}{dv} \in \mathbb{R}$ is the variation of the out-of-plane force to the wind speed and $r_l \in \mathbb{R}$ is the blade length. Notice that this is similar to the wind-induced blade disturbance M_i^{d}, discussed in Sect. 2.4. The benefits of the blade effective wind speed is that the wind disturbance at higher frequencies can be retained compared to the simplified wind model (3.23). More discussions are available in Chap. 7.

3.4.2 Feed-Forward Design Methods in Wind Turbine

In this section, the related work on wind turbine feed-forward control is presented based on their design methods.

3.4.2.1 Model Inversion-Based Control

Model inversion-based control is one of the simplest forms of the feed-forward control. Consider a linear model as follows:

$$y(s) = G(s)u(s) + G_{\mathrm{d}}(s)d(s), \tag{3.26}$$

where y, u, d are the output, input and measured disturbance, whilst G, G_{d} are the transfer functions of the plant and disturbance model. Assuming the system model G is invertible, a simple but effective feed-forward control law is as follows:

$$u(s) = -G(s)^{-1}G_{\mathrm{d}}(s)d(s). \tag{3.27}$$

This model inversion-based method is popular in wind turbine control. For example, Schlipf and Kuhn (2008), Wang, Yue, Bao, and Leithead (2015) and Bao, Wang, Yue, and Leithead (2015) used the model inversion-based design addressing the rotor speed regulation problem, whilst Schlipf et al. (2010) proposed a model inversion-based feed-forward controller for alleviating the blade loads. Similarly, a gain-scheduling model inversion-based controller was also proposed by Dunne, Pao, and Wright (2010) and Dunne et al. (2011) for turbine load reduction. The drawbacks of this method are that the constraints are not taken into account and also the plant model G is not necessarily invertible.

3.4.2.2 Non-model Based Control

In recent years, a non-model based approach was proposed by Dunne, Pao, Wright, Jonkman, Kelley, and Simley (2011) and Dunne et al. (2011) for attenuating blade loads, where the feed-forward controller was formed heuristically by shaping the pre-filter. Similarly, Wang, Johnson, Wright, and Wright (2012) and Wang (2013) extended the non-model based approach to an adaptive design where the feed-forward control law was updated on-line based on the wind conditions.

3.4.2.3 Model-Based Control

Some studies adopt optimisation-based methods to design feed-forward controllers. For example, $\mathcal{H}_2$ design by Dunne and Pao (2013, 2016) and Dunne (2016) for regulating the rotor speed, $\mathcal{H}_\infty$ design by Laks et al. (2009) for blade load attenuation. However, as reported in Laks et al. (2009), the use of the approaching wind measurements might potentially drive the pitch actuators to operate around their limits, especially for the IPC, thus, a model predictive control design is motivated.

3.4.2.4 Model Predictive Control

MPC in Rotor Speed Regulation

One of the most popular types of optimisation-based control methods in wind turbines is model predictive control. This is because, as demonstrated by Henriksen (2007, 2010), Henriksen, Hansen, and Poulsen (2012), the optimisation-based control law, that incorporates the *constraint knowledge* on pitch actuator and generator torque input, easily outperformed the proportional-integral (PI) controller with integrator anti-windup. Similarly, Soliman, Malik, and Westwick (2010), Kumar and Stol (2009) and Hur and Leithead (2016) adopted a gain-scheduling linear MPC approach, where the control laws were calculated based on a range of wind turbine models derived from both below and above-rated wind conditions. Furthermore, some studies, for example, Bottasso, Croce, and Savini (2007) and Dang, Wang, and Cai (2008) proposed nonlinear MPC controllers in order to overcome the nonlinearity between these operating wind conditions.

Another key reason for the popularity of MPC in wind turbines is the *systematic incorporation of the future wind information*. As demonstrated by Körber and King (2010), the linear MPC control law, that takes into account the perfect upcoming wind information, achieved better rotor speed regulation than a baseline PI controller. In a follow-up study by Körber and King (2011), a nonlinear MPC approach was adopted assuming the preview knowledge was known perfectly in advance. Later, Schlipf, Schlipf, and Kühn (2013) extended the study of nonlinear MPC to a scenario where the measurement errors from the LIDAR system were included.

MPC in Blade Load Reduction

Model predictive control approach was adopted by Laks et al. (2011) and Laks (2013), where the constraints on pitch rate and pitch magnitudes as well as the upcoming wind measurements from the LIDAR systems were included in the online optimisation. A similar study by Spencer et al. (2013), Soltani, Wisniewski, Brath, and Boyd (2011) and Madsen, Filso, and Soltani (2012) also demonstrated that the MPC design yielded good performance when handling the approaching wind information. To counter the model variations in different operating conditions, Mirzaei (2012), Mirzaei, Henriksen, Poulsen, Niemann, and Hansen (2012), Mirzaei and Soltani (2013) and Mirzaei, Soltani, Poulsen, and Niemann (2013) proposed a gain-scheduling MPC design and also an algorithm to correct the preview time difference between the estimated and actual wind measurements.

Discussions

Despite recent development of some MPC theories, such as tube-based MPC (e.g. Cannon, Kouvaritakis, Rakovic, & Cheng, 2011a; Chisci, Rossiter, & Zappa, 2001; Kouvaritakis, Cannon, Raković, & Cheng, 2010; Mayne, Seron, & Rakovic, 2005) and Laguerre function-based MPC (e.g. Rossiter & Wang, 2008; Valencia-Palomo, Rossiter, Jones, Gondhalekar, & Khan, 2011; Wang, 2004), the application of these advanced MPC techniques for wind turbines is still in its infancy. Körber (2014) and Körber and King (2013) addressed the turbine speed regulation problem with perfect advance wind knowledge by a robust MPC approach where the constraints were formulated including all the possible violations based on the prior knowledge of the disturbance model. In addition, one non-preview robust MPC approach in wind turbines was studied by Evans, Cannon, and Kouvaritakis (2015) for tower load damping. Furthermore, Adegas, Wisniewski, and Sloth Larsen (2013) employed a Laguerre function-based MPC in wind turbines where such parametrisations of degrees-of-freedom simplified the rotor speed controller design which could be achieved by placement of closed-loop poles.

3.4.3 Wind Measurement Quality

So far, it is assumed that the wind field is accurately measured by the LIDAR devices and the wind turbulence does not evolve between the measurement point and the turbine, known as Taylor's frozen turbulence hypothesis (Taylor, 1938). However, in reality, the turbulence evolves as the wind field moves towards the wind turbine from the point of measurements. Therefore, Bossanyi and Garrad Hassan (2012) addressed these issues and conducted studies of preview control under more realistic LIDAR measurements. Recent studies by (Wang et al., 2015; Bao et al., 2015) proposed a feed-forward controller based on a system inversion technique that uses simulated LIDAR measurements including the evolving turbulence.

Nonetheless, since wind-field estimation is a non-trivial problem and the main focus of this thesis is on preview control design, thus, details of recent developments in the wind-field estimation techniques are not included in this thesis and some excellent sources can be found in Raach, Schlipf, Haizmann, and Cheng (2014), Towers and Jones (2016), Schlipf et al. (2010), Bossanyi and Garrad Hassan (2012), Simley, Pao, Frehlich, Jonkman, and Kelley (2014), Simley and Pao (2012, 2013), Simley et al. (2011, 2016), Simley (2015) and the references therein.

3.4.4 *Summary*

This section presented the recent development in wind field modelling. The key features of various modelling methods were discussed. In addition, control design methods for feed-forward wind turbine control were reviewed. In particular, the related work regarding the MPC in wind turbine was discussed in Sect. 3.4.2.4. To sum up, this section laid an important foundation for Chap. 7, where the upcoming wind information is used for feed-forward control.

3.5 Summary

The backgrounds of three topics were covered in this chapter. Firstly, literature regarding individual pitch control was discussed, that sets the scene for Chap. 4. Secondly, discussions in Sect. 3.3 highlighted the recent development in feed-forward model predictive control. And lastly, related work regarding many aspects of the feed-forward control in wind turbine provided an important insight to the reader for understanding the topics in Chap. 7.

In the coming chapter, the performance similarity of different classes of the IPCs is investigated, that is heavily based upon the related works presented in Sect. 3.2.

References

Adegas, F. D., Wisniewski, R., & Sloth Larsen, L. F. (2013). Gain-scheduled model predictive control of wind turbines using Laguerre functions. In *2013 American Control Conference* (pp. 653–658). New York: IEEE.

Bao, J., Wang, M., Yue, H., & Leithead, W. (2015). Pseudo-lidar data analysis and feed-forward wind turbine control design. In *IFAC ADCHEM 2015* (pp. 483–488).

Barlas, T., & van Kuik, G. (2010). Review of state of the art in smart rotor control research for wind turbines. *Progress in Aerospace Sciences*, *46*(1), 1–27.

Bossanyi, E. A., & Garrad Hassan, G. L. (2012). Un-freezing the turbulence: Improved wind field modelling for investigating Lidar-assisted wind turbine control. In *Proceedings of EWEA 2012*.

Bossanyi, E. A., & Wright, A. (2009). Field testing of individual pitch control on the NREL CART-2 wind turbine. In *European Wind Energy Conference*.

Bossanyi, E., Savini, B., Iribas, M., Hau, M., Fischer, B., Schlipf, D., ..., Carcingiu (2012). Advanced controller research for multi-MW wind turbines in the UPWIND project. *Wind Energy*, *15*(1), 119–145.

Bossanyi, E. A. (2003). Individual blade pitch control for load reduction. *Wind Energy*, *6*(2), 119–128.

Bossanyi, E. A. (2005). Further load reductions with individual pitch control. *Wind Energy*, *8*(4), 481–485.

Bossanyi, E. A., Fleming, P. A., & Wright, A. D. (2013). Validation of individual pitch control by field tests on two- and three-bladed wind turbines. *IEEE Transactions on Control Systems Technology*, *21*(4), 1067–1078.

Bottasso, C., Croce, A., Riboldi, C., & Nam, Y. (2013). Multi-layer control architecture for the reduction of deterministic and non-deterministic loads on wind turbines. *Renewable Energy, 51*, 159–169.

Bottasso, C. L., Croce, A., & Savini, B. (2007). Performance comparison of control schemes for variable-speed wind turbines. *Journal of Physics: Conference Series, 75*, 012079.

Cannon, M., Kouvaritakis, B., Rakovic, S. V., & Cheng, Q. (2011). Stochastic tubes in model predictive control with probabilistic constraints. *IEEE Transactions on Automatic Control, 56*(1), 194–200.

Carrasco, D. S., & Goodwin, G. C. (2011a). Feedforward model predictive control. *Annual Reviews in Control, 35*(2), 199–206.

Carrasco, D. S., & Goodwin, G. C. (2011b). Preview and feedforward in model predictive control: A preliminary robustness analysis. In *IFAC World Congress* (Vol. 18, pp. 185–190).

Caselitz, P., Kleinkauf, W., Kruger, T., Petschenka, J., Reichardt, M., & Storzel, K. (1997). Reduction of fatigue loads on wind energy converters by advanced control methods. In *European Wind Energy Conference* (pp. 555–558). Dublin.

Chisci, L., Rossiter, J., & Zappa, G. (2001). Systems with persistent disturbances: Predictive control with restricted constraints. *Automatica, 37*, 1019–1028.

Clarke, D., Mohtadi, C., & Tuffs, P. (1987a). Generalized predictive control—Part II. The basic algorithm. *Automatica, 23*(2), 137–148.

Clarke, D., Mohtadi, C., & Tuffs, P. (1987b). Generalized predictive control—Part I. The basic algorithm. *Automatica, 23*(2), 137–148.

Coleman, R. P., & Feingold, A. M. (1957). Theory of self-excited mechanical oscillations of helicopter rotors with hinged blades. *National Advisory Committee for Aeronautics*, Report 1351.

Dang, D., Wang, Y., & Cai, W. (2008). Nonlinear model predictive control (NMPC) of fixed pitch variable speed wind turbine. In *Sustainable energy technologies* (Vol. 2, pp. 29–33).

Duesterhoeft, W. C., Schulz, M. W., & Clarke, E. (1951). Determination of instantaneous currents and voltages by means of alpha, beta, and zero components. *Transactions of the American Institute of Electrical Engineers, 70*(2), 1248–1255.

Dunne, F. (2016). *Optimizing blade pitch control of wind turbines with preview measurements of the wind by* (Ph.D. thesis). University of Colorado.

Dunne, F., & Pao, L. Y. (2016). Optimal blade pitch control with realistic preview wind measurements. *Wind Energy*.

Dunne, F., Pao, L., & Wright, A. (2010). Combining standard feedback controllers with feedforward blade pitch control for load mitigation in wind turbines. In *48th AIAA Aerospace Sciences Meeting* (pp. 1–18).

Dunne, F., & Pao, L. Y. (2013). Benefit of wind turbine preview control as a function of measurement coherence and preview time. *American Control Conference (ACC), 2013*, 647–652.

Dunne, F., Pao, L. Y., Wright, A. D., Jonkman, B., & Kelley, N. (2011). Adding feedforward blade pitch control to standard feedback controllers for load mitigation in wind turbines. *Mechatronics, 21*, 682–690.

Dunne, F., Pao, L., Wright, A., Jonkman, B., Kelley, N., & Simley, E. (2011). Adding feedforward blade pitch control for load mitigation in wind turbines: Non-causal series expansion, preview control, and optimized FIR filter methods. In *49th AIAA Aerospace Sciences Meeting Including the New Horizons Forum and Aerospace Exposition* (pp. 1–17). American Institute of Aeronautics and Astronautics: Reston, VA.

Evans, M. A., Cannon, M., & Kouvaritakis, B. (2015). Robust MPC tower damping for variable speed wind turbines. *IEEE Transactions on Control Systems Technology, 23*(1), 290–296.

Geyler, M., & Caselitz, P. (2007). Individual blade pitch control design for load reduction on large wind turbines. In *European Wind Energy Conference (EWEC 2007)*.

Geyler, M., & Caselitz, P. (2008). Robust multivariable pitch control design for load reduction on large wind turbines. *Journal of Solar Energy Engineering, 130*(3), 031014.

Goodwin, G., Carrasco, D., & Mayne, D. (2011). Preview and feedforward in model predictive control: Conceptual and design issues. In *IFAC World Congress* (pp. 5555–5560).

Han, Y., & Leithead, W. E. (2014). Combined wind turbine fatigue and ultimate load reduction by individual blade control. In *Proceedings of the Science of Making Torque from Wind* (Vol. 524, p. 012062).

Harris, M., Hand, M., & Wright, A. (2005). *Lidar for turbine control* (Technical Report).

Henriksen, L. C. (2007). *Model predictive control for wind turbines* (Master's thesis). Technical University of Denmark.

Henriksen, L. C. (2010). *Model predictive control for wind turbines* (Ph.D. thesis). Technical University of Denmark.

Henriksen, L., Hansen, M., & Poulsen, N. (2012). Wind turbine control with constraint handling: A model predictive control approach. *IET Control Theory & Applications*, *6*(11), 1722.

Hur, S., & Leithead, W. E. (2016). Model predictive and linear quadratic Gaussian control of a wind turbine. *Optimal Control Applications and Methods*, *143*.

Jasniewicz, B., & Geyler, M. (2011). Wind turbine modelling and identification for control system applications. In *EWEA*. Brussels, Belgium.

Jelavic, M., Petrovic, V., & Peric, N. (2008). Individual pitch control of wind turbine based on loads estimation. In *Annual Conference of IEEE Industrial Electronics*.

Jonkman, J., & Buhl Jr, M. (2005). *FAST user's guide* (Technical Report). National Renewable Energy Laboratory (NREL).

Kodama, N., Matsuzaka, T., Tuchiya, K., & Arinaga, S. (1999). Power variation control of a wind generator by using feed forward control. *Renewable Energy*, *16*, 847–850.

Körber, A. (2014). *Extreme and fatigue load reducing control for wind turbines : A model predictive control approach using robust state constraints* (Ph.D. thesis).

Körber, A., & King, R. (2010). Model predictive control for wind turbines. In *Proceedings of European Wind Energy Conference* (pp. 3–9).

Körber, A., & King, R. (2011). Nonlinear model predictive control for wind turbines. In *EWEA* (pp. 1–6).

Körber, A., & King, R. (2013). Combined feedback-feedforward control of wind turbines using state-constrained model predictive control. *IEEE Transactions on Control Systems Technology*, *21*(4), 1117–1128.

Kouvaritakis, B., Cannon, M., Raković, S. V., & Cheng, Q. (2010). Explicit use of probabilistic distributions in linear predictive control. *Automatica*, *46*(10), 1719–1724.

Kouvaritakis, B., Rossiter, J., & Cannon, M. (1998). Linear quadratic feasible predictive control. *Automatica*, *34*(12), 1583–1592.

Kumar, A., & Stol, K. (2009). Scheduled model predictive control of a wind turbine. In *Proceedings of AIAA/ASME Wind Energy Symposium*.

Lackner, M. A., & van Kuik, G. (2010). A comparison of smart rotor control approaches using trailing edge aps and individual pitch control. *Wind Energy*, *13*(2–3), 117–134.

Laks, J. H. (2013). *Preview scheduled model predictive control for horizontal axis wind turbines* (Ph.D. thesis). University of Colorado.

Laks, J., Pao, L., Simley, E., Wright, A., Kelley, N., & Jonkman, B. (2011). Model predictive control using preview measurements from LIDAR. In *49th AIAA*. Reston, VA: American Institute of Aeronautics and Astronautics.

Laks, J., Pao, L. Y., Wright, A., Kelley, N., & Jonkman, B. (2010). Blade pitch control with preview wind measurements. In *Proceedings of 48th AIAA*.

Laks, J., Pao, L., & Wright, A. (2009a). Combined feed-forward/feedback control of wind turbines to reduce blade flap bending moments. In *AIAA/ASME Wind Energy Symposium*. Orlando, FL.

Laks, J., Pao, L., Wright, A., Kelley, N., & Jonkman, B. (2011). The use of preview wind measurements for blade pitch control. *Mechatronics*, *21*(4), 668–681.

Leithead, W., Dominguez, S., & Spruce, C. (2004). Analysis of tower/blade interaction in the cancellation of the tower fore-aft mode via control. In *EWEC*.

Leithead, W., Neilson, V., & Dominguez, S. (2009a). Alleviation of unbalanced rotor loads by single blade controllers. In *European Wind Energy Conference*.

Leithead, W., Neilson, V., Dominguez, S., & Dutka, A. (2009b). A novel approach to structural load control using intelligent actuators. In *2009 17th Mediterranean Conference on Control and Automation* (pp. 1257–1262). New York: IEEE.

Lu, Q., Bowyer, R., & Jones, B. (2015). Analysis and design of Coleman transformbased individual pitch controllers for wind-turbine load reduction. *Wind Energy*, *18*(8), 1451–1468.

Madsen, M. B., Filso, J., & Soltani, M. (2012). Preview-based asymmetric load reduction of wind turbines [Master's thesis]. In *2012 IEEE International Conference on Control Applications* (pp. 1424–1429). New York: IEEE.

Mayne, D. Q., Seron, M. M., & Rakovic, S. V. (2005). Robust model predictive control of constrained linear systems with bounded disturbances. *Automatica*, *41*(2), 219–224.

Mirzaei, M. (2012). *Wind turbine control: A robust model based approach* (Ph.D. thesis). Technical University of Denmark.

Mirzaei, M., & Soltani, M. (2013). Model predictive control of wind turbines using uncertain LIDAR measurements. In *Proceedings of the American Control Conference*, (pp. 2235–2240).

Mirzaei, M., Henriksen, L. C., Poulsen, N. K., Niemann, H. H., & Hansen, M. H. (2012). Individual pitch control using LIDAR measurements. In *Proceedings of IEEE International Conference on Control Applications* (pp. 1646–1651).

Mirzaei, M., Soltani, M., Poulsen, N. K., & Niemann, H. H. (2013). An MPC approach to individual pitch control of wind turbines using uncertain LIDAR measurements. In *Proceedings of ECC* (pp. 2238–2243).

Park, R. H. (1929). Two-reaction theory of synchronous machines generalized method of analysis-part I. *Transactions of the American Institute of Electrical Engineers*, *48*(3), 716–727.

Plumley, C., Leithead, W., Jamieson, P., Bossanyi, E., & Graham, M. (2014). Comparison of individual pitch and smart rotor control strategies for load reduction. In *Proceedings of the Science of Making Torque from Wind* (Vol. 524, p. 012054).

Raach, S., Schlipf, D., Haizmann, F., & Cheng, P. W. (2014). Three dimensional dynamic model based wind field reconstruction from lidar data. In *The science of making torque from wind*.

Rossiter, J. (2015). *Video lectures on modelling, analysis, control and MPC*.

Rossiter, J. A., & Grinnell, B. G. (1995). Extended input horizon generalized predictive control—A GPC algorithm with better tracking. In *3rd IEEE Symposium on New Directions in Control and Automation*. Cyprus.

Rossiter, J. A., & Grinnell, B. G. (1996). Improving the tracking of generalized predictive control controllers. In *Proceedings of IMECHE*.

Rossiter, J. A., & Valencia-Palomo, G. (2009). Feed forward design in MPC. In *European Control Conference*.

Rossiter, J. A., & Wang, L. (2008). Exploiting Laguerre functions to improve the feasibility/performance compromise in MPC. In *2008 47th IEEE Conference on Decision and Control* (pp. 4737–4742). New York: IEEE.

Schlipf, D. (2013). 9 Lidars and wind turbine control—Part 1.

Schlipf, D., & Kuhn, M. (2008). Prospects of a collective pitch control by means of predictive. In *Proceedings of the 9th German Wind Energy Conference*.

Schlipf, D., Fischer, T., & Carcangiu, C. (2010). Load analysis of look-ahead collective pitch control using LIDAR. In *Proceedings of 10th German Wind Energy Conference*.

Schlipf, D., Schuler, S., Grau, P., & Martin, K. (2010). Look-ahead cyclic pitch control using LIDAR. In *The science of making torque from wind*.

Schlipf, D., Schlipf, D. J., & Kühn, M. (2013). Nonlinear model predictive control of wind turbines using LIDAR. *Wind Energy*, *16*(7), 1107–1129.

Scokaert, P. O. M., & Rawlings, J. B. (1998). Constrained linear quadratic regulation. *IEEE Transactions on Automatic Control*, *43*(8), 1163–1169.

Selvam, K. (2007). *Individual pitch control for large scale wind turbines multivariable control approach* (Technical Report). ECN.

Selvam, K., Kanev, S., van Wingerden, J. W., van Engelen, T., & Verhaegen, M., (2009). Feedback-feedforward individual pitch control for wind turbine load reduction. *International Journal of Robust and Nonlinear Control*, *19*(1), 72–91.

Shan, M., & Adelt, S. (2013). Field testing and practical aspects of load reducing pitch control systems for a 5 MW offshore wind turbine. In *EWEA*.

Simley, E. J. (2015). *Wind speed preview measurement and estimation for feedforward control of wind turbines* (Ph.D. thesis). University of Colorado.

Simley, E., & Pao, L. Y. (2012). *LIDAR wind speed measurements of evolving wind fields* (Technical Report). NREL.

Simley, E., & Pao, L. (2013). Correlation between rotating LIDAR measurements and blade effective wind speed. In *AIAA Aerospace Sciences Meeting*, (pp. 1–15).

Simley, E., Dunne, F., Laks, J., & Pao, L. Y. (2013). *LIDARS and wind turbine control Part 2* (Technical Report). DTU Wind Energy.

Simley, E., Pao, L., & Frehlich, R. (2011). Analysis of wind speed measurements using continuous wave LIDAR for wind turbine control. In *Proceedings of 49th AIAA*.

Simley, E., Pao, L. Y., Frehlich, R., Jonkman, B., & Kelley, N. (2014). Analysis of light detection and ranging wind speed measurements for wind turbine control. *Wind Energy* (February 2013), 413–433.

Simley, E., Angelou, N., Mikkelsen, T., Sjöholm, M., Mann, J., & Pao, L. Y. (2016). Characterization of wind velocities in the upstream induction zone of a wind turbine using scanning continuous-wave lidars. *Journal of Renewable and Sustainable Energy*, *8*(1), 013301.

Soliman, M., Malik, O., & Westwick, D. (2010). Multiple model MIMO predictive control for variable speed variable pitch wind turbines. In *Proceedings of ACC* (pp. 2778–2784).

Soltani, M., Wisniewski, R., Brath, P., & Boyd, S. (2011). Load reduction of wind turbines using receding horizon control. In *2011 IEEE International Conference on Control Applications (CCA)* (pp. 852–857). New York: IEEE.

Spencer, M. D., Stol, K. A., Unsworth, C. P., Cater, J. E., & Norris, S. E. (2013). Model predictive control of a wind turbine using short-term wind field predictions. *Wind Energy*, *16*(3), 417–434.

Stol, K. A., Zhao, W., & Wright, A. D. (2006). Individual blade pitch control for the controls advanced research turbine (CART). *Journal of Solar Energy Engineering*, *128*(4), 498.

Taylor, G. I. (1938). The spectrum of turbulence. *Proceedings of the Royal Society A: Mathematical, Physical and Engineering Sciences*, *164*(919), 476–490.

Towers, P., & Jones, B. L. (2016). Real-time wind field reconstruction from LIDAR measurements using a dynamic wind model and state estimation. *Wind Energy*, *19*(1), 133–150.

Valencia-Palomo, G., Rossiter, J. A., & López-Estrada, F. R. (2014). Improving the feedforward compensator in predictive control for setpoint tracking. *ISA Transactions*, *53*(3), 755–66.

Valencia-Palomo, G., Rossiter, J., Jones, C., Gondhalekar, R., & Khan, B. (2011). Alternative parameterisations for predictive control: How and why? In *Proceedings of the 2011 American Control Conference*, (pp. 5175–5180).

Vali, M., van Wingerden, J.-W., & Kuhn, M. (2016). Optimal multivariable individual pitch control for load reduction of large wind turbines. In *2016 American Control Conference (ACC)* (pp. 3163–3169). New York: IEEE.

van Engelen, T. G. (2006). Design model and load reduction assessment for multirotational mode individual pitch control (higher harmonics control). In *Proceedings of European Wind Energy Conference*.

van Engelen, T., & Braam, H. (2004). *TURBU offshore: Implementation* (Technical Report). ECN Wind Energy.

van Engelen, T. G., & van der Hooft, E. L. (2005). *Individual pitch control inventory* (Technical Report). ECN.

Wang, N. (2013). *LIDAR-assisted feedforward and feedback control design for wind turbine tower load mitigation and power capture enhancement* (Unpublished doctoral dissertation).

Wang, L., & Rossiter, J. A. (2008). Disturbance rejection and set-point tracking of sinusoidal signals using generalized predictive control. In *Proceedings of CDC* (Vol. 1, pp. 4079–4084).

Wang, N., Johnson, K. E., Wright, A. D., & Wright., A. D., (2012). FX-RLS-based feedforward control for LIDAR-enabled wind turbine load mitigation. *IEEE Transactions on Control Systems Technology*, *20*(5), 1212–1222.

Wang, M., Yue, H., Bao, J., & Leithead, W. E. (2015). LIDAR-based wind speed modelling and control system design. In *2015 21st International Conference on Automation and Computing (ICAC)* (pp. 1–6). New York: IEEE.

Wang, L. (2004). Discrete model predictive controller design using Laguerre functions. *Journal of Process Control*, *14*, 131–142.

Yi, H., & Leithead, W. (2012). Alleviation of extreme blade loads by individual blade control during normal wind turbine operation. In *Proceedings of EWEA*.

Zhang, Y., Cheng, M., & Chen, Z. (2013). Proportional resonant individual pitch control for mitigation of wind turbines loads. *IET Renewable Power Generation*, *7*(3), 191–200.

Chapter 4
Performance Similarities Between Individual Pitch Control Strategies

4.1 Overview and Contribution

The use of blade individual pitch control (IPC) offers a means of reducing the harmful turbine structural loads that arise from the uneven and unsteady forcing from the oncoming wind. In recent years two different and competing IPC techniques have emerged that are characterised by the specific loads that they are primarily designed to attenuate. In the first instance, methodologies such as single-blade control and Clarke Transform-based control have been developed to reduce the unsteady loads on the rotating blades, whilst tilt-yaw control and its many variants instead target load reductions in the non rotating turbine structures, such as the tower and main bearing. Given the seeming disparities between these controllers, the aim of this chapter is to investigate and understand the fundamental performance similarities that exist between them and hence unify research in this area. And a significant new result in this chapter shows that single-blade controllers are equivalent to a particular class of tilt-yaw controller, which itself is equivalent to Clarke Transform-based control. This means that three architecturally dissimilar IPC controllers exist that yield exactly the same performance in terms of load reductions on fixed and rotating turbine structures. This chapter further demonstrates this outcome by presenting results obtained from high-fidelity closed-loop turbine simulations.

This chapter is structured as follows: Sect. 4.2 presents an introduction and motivation of the chapter. Section 4.3 defines the three different IPCs under comparison. These are a Coleman Transform-based controller, a Clarke Transform-based controller, and a single-blade controller. In Sect. 4.4, the equivalence between these IPCs is established. Specifically, this chapter shows; (i) that a single-blade controller is equivalent to a Coleman Transform-based controller with a particular structure; (ii) that this Coleman Transform-based controller is equivalent to a Clarke Transform-based controller; and (iii) that all three IPCs yield identical performance, as quantified by the robust stability margin. In Sect. 4.5, this equivalence is demonstrated by performing separate closed-loop simulations upon a high-fidelity wind turbine model, followed by a discussion of the results. This chapter is concluded in Sect. 4.6.

W. H. (Alan) Lio, *Blade-Pitch Control for Wind Turbine Load Reductions*, Springer Theses, https://doi.org/10.1007/978-3-319-75532-8_4

4.2 Introduction

The ability possessed by most modern wind turbine generators to actively control the pitch of each blade offers the potential to reduce the unsteady loads that arise form a number of sources, such as wind-shear, tower shadow, yaw misalignment and turbulence within the atmospheric boundary layer (Barlas & van Kuik, 2010). Such loads are a known source of the structural fatigue damage that can reduce the operational lifetime of a turbine, ultimately increasing the cost of wind energy to the end user. As a consequence, a growing body of research has emerged in recent years, seeking to establish the best way of designing individual pitch control (IPC) systems. Typically, and for reasons of simplicity of implementation favoured by the industry, IPCs are designed separately from a collective pitch control (CPC) system, whose role is to regulate the rotor speed in above-rated wind conditions by collectively adjusting the pitch angle of each blade by the same amount (Muljadi, 2001; Pao & Johnson, 2009). The IPC provides an additional pitch angle demand signal, typically in response to measurements of the flap-wise blade root bending moments, in order to attenuate the effects of unsteady spatio-temporal rotor loads.

Of the many IPC strategies that have been published in recent years, most can be grouped into two distinct classes, characterised by the specific turbine loads they are primarily designed to attenuate. The first and most popular branch of IPC targets load reductions on the non-rotating turbine structures, such as the tower, nacelle and main bearing. A coordinate transformation is employed to refer sensing and actuation signals in the rotating frame of reference to a non-rotating reference frame. The most commonly employed transformation in this respect is the Coleman Transform. As noted by Lu, Bowyer, and Jones (2015), this transformation emerged from the area of helicopter rotor control (Coleman & Feingold, 1957), and is widely employed in the fields of power conversion and electrical machines under the guise of the direct-quadrature-zero (dq0) transform (Vas, 1992). Use of the Coleman Transform to address the IPC problem was adopted by Bossanyi (2003a) and van Engelen and van der Hooft (2005) in order to project blade loads onto the non-rotating and orthogonal turbine tilt and yaw axes. Subsequent IPC design then attenuates the tilt and yaw referred loads, with such designs sometimes referred to as 'tilt-yaw' controllers. These produce tilt and yaw referred pitch demand signals which are projected back to the rotating frame of reference via the inverse Coleman Transform. The attractive feature of the Coleman Transform is that it transforms an otherwise time periodic system into one that is time-invariant by projecting the system inputs and outputs in the rotational frame of reference onto stationary tilt and yaw axes. If the turbine dynamics are linear, or can be approximated as such, then conventional tools of linear and time-invariant (LTI) control system design can further be applied to design controllers to attenuate the unsteady loads upon the non-rotating turbine structures. This is the main reason why the majority of IPC studies have employed the Coleman Transform (Bossanyi, 2003a, 2005; Bossanyi & Wright, 2009; Engels,

Subhani, Zafar, & Savenije, 2014; Geyler & Caselitz, 2008; Lackner & van Kuik, 2010; Lu et al., 2015; Plumley et al., 2014; Selvam, Kanev, van Wingerden, van Engelen, & Verhaegen, 2009; Stol, Moll, Bir, & Namik, 2009; van Engelen, 2006; van Engelen & van der Hooft, 2005).

The second branch of IPC targets load reductions upon the rotating turbine structures, primarily the blades. Single-blade control (Leithead, Neilson, & Dominguez, 2009; Leithead, Neilson, Dominguez, & Dutka, 2009), later termed individual blade control (Han & Leithead, 2014), equips each blade with its own controller that actuates in response to local blade load measurements. The overall IPC controller is thus formed from three identical single-input-single-output (SISO) controllers acting independently from one another. Although conceptually simple, there is redundancy in the sense that three separate SISO controllers are not necessary to design an IPC controller. Recently, Zhang, Cheng, & Chen, (2013) showed it was possible to use just two identical SISO controllers, pre and post-compensated by the Clarke Transform (Vas, 1992) and its inverse to yield good blade-load reductions. This form of blade load IPC was termed proportional-resonant control by these authors. It is interesting to note that the Clarke Transform, also known as the $\alpha\beta\gamma$ Transform, is conceptually similar to the Coleman Transform in the sense that both transforms perform projections onto a set of orthogonal axes. However, whereas the Coleman Transform performs a projection onto a set of axes that are rotating with respect to the turbine blades, the Clarke Transform performs a projection onto a set of axes that are stationary with respect to the blades. One immediate implication of this, as noted by Zhang et al. (2013), is that the Clarke Transform-based IPC does not require a measurement of the rotor azimuth angle, unlike IPC based upon the Coleman Transform. The same benefit also holds for single-blade control.

Given this range of IPC techniques, it is natural to attempt to understand under what conditions these different controllers yield similar performance, in terms of load reductions. However, this is not as straightforward as it may seem. A detailed literature review was presented in Sect. 3.2 in Chap. 3. The fashion in which load reductions about the tilt and yaw axes correspond to reductions in blade loads is somewhat complicated by virtue of the frequency shifting effects of the Coleman Transform (Lu et al., 2015). As demonstrated in Sect. 2.9 in Chap. 2, wind turbine loads predominantly exist at the harmonics of the blade rotational frequency (Barlas & van Kuik, 2010). For three-bladed turbines, the blade loads are concentrated at integer multiples of the once per revolution (1p) blade frequency, resulting in non-rotating loads at adjacent harmonics to the nearest 3p frequency (Zhang et al., 2013). For example, 1p blade loads map to static (0p) loads in the tilt and yaw frame of reference, whilst 3p non-rotating structural loads are split into 2p and 4p blade loads. It is this frequency shifting of loads that makes IPC comparisons difficult, and understanding this problem forms the essence of this chapter.

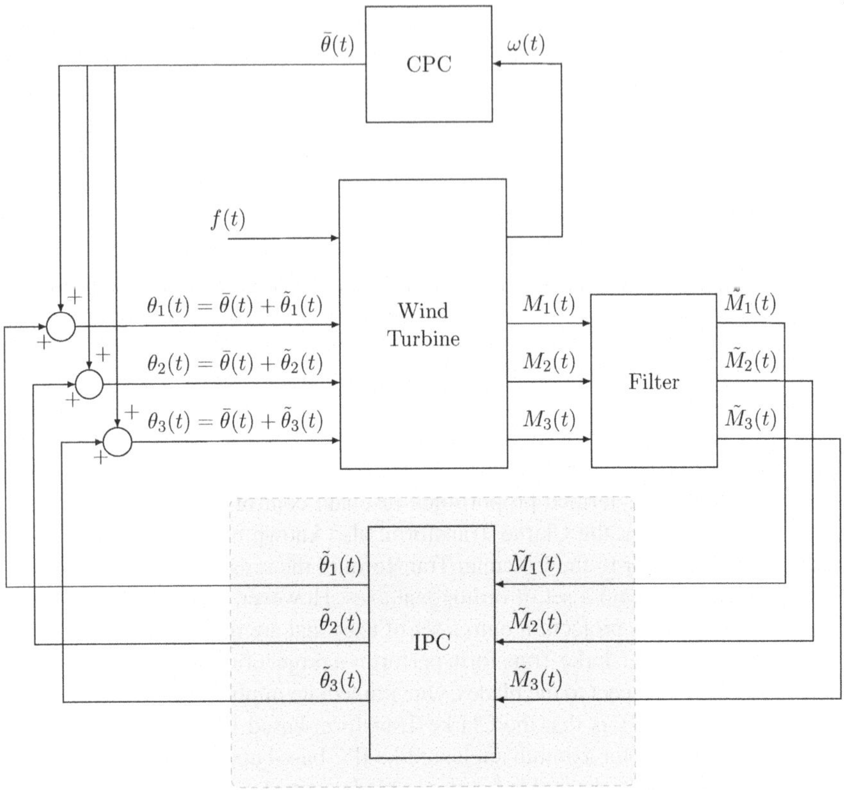

Fig. 4.1 System architecture of a wind turbine, combining collective pitch control (CPC) and individual pitch control (IPC). The CPC regulates rotor speed while the IPC (shaded) attenuates perturbations in the flap-wise root bending moments on each blade. Additional inputs to the turbine such as wind loading and generator torque are accounted for in the term $f(t)$

4.3 Individual Pitch Control

This section presents an overview summary of various types of IPCs. Some details in this section may appear in the early chapters. However, It is helpful for the reader because the notations of this chapter are properly defined in this section.

A typical wind turbine control systems architecture for above-rated conditions is shown in Fig. 4.1. The CPC regulates the rotor speed $\omega(t)$ by adjusting the collective pitch angle $\bar{\theta}(t)$. To isolate the action of IPC from that of CPC, it is convenient to define the pitch angles and blade moments as follows:

$$\begin{bmatrix}\theta_1(t)\\ \theta_2(t)\\ \theta_3(t)\end{bmatrix} := \begin{bmatrix}\bar{\theta}(t)+\tilde{\theta}_1(t)\\ \bar{\theta}(t)+\tilde{\theta}_2(t)\\ \bar{\theta}(t)+\tilde{\theta}_3(t)\end{bmatrix}, \quad \begin{bmatrix}M_1(t)\\ M_2(t)\\ M_3(t)\end{bmatrix} := \begin{bmatrix}\bar{M}(t)+\tilde{M}_1(t)\\ \bar{M}(t)+\tilde{M}_2(t)\\ \bar{M}(t)+\tilde{M}_3(t)\end{bmatrix}, \tag{4.1}$$

where $\tilde{\theta}_{1,2,3}(t)$ represent the perturbations in blade pitch angle demand from the collective pitch signal, whilst $\tilde{M}_{1,2,3}(t)$ are the perturbations in flap-wise blade bending moments, obtained by filtering out the mean moment $\bar{M}(t)$ from the measurements $M_{1,2,3}(t)$. This filtering is important in order to help decouple the IPC from the CPC. For each blade, the relationship between perturbation input $\tilde{\theta}_i$ and output $\tilde{M}_i$, for $i \in \{1, 2, 3\}$ can be modelled by a transfer function $G \in \mathcal{R}$, obtaining by linearising the turbine dynamics around the rated rotor speed ω_0. A typical blade transfer function, as used by Lu et al. (2015) for example, is as follows:

$$G(s) := G_{\mathrm{a}}(s)G_{\mathrm{b}}(s)G_{\mathrm{bp}}(s), \tag{4.2a}$$

where $G_{\mathrm{a}}, G_{\mathrm{b}} \in \mathcal{R}$ describe the dynamics of the pitch actuator and the blade, respectively, whilst $G_{\mathrm{bp}} \in \mathcal{R}$ is a band-pass filter that is included in order to remove the low frequency component of the flap-wise blade root bending moment signal from strain-gauge sensors, as well as high frequency noise. Basic models for each of these transfer functions are as follows:

$$G_{\mathrm{a}}(s) := \frac{1}{1+\tau s}, \tag{4.2b}$$

$$G_{\mathrm{b}}(s) := \frac{dM_{\mathrm{flap}}}{d\theta}\frac{(2\pi f_{\mathrm{b}})^2}{s^2 + D_{\mathrm{b}}2\pi f_{\mathrm{b}}s + (2\pi f_{\mathrm{b}})^2}, \tag{4.2c}$$

$$G_{\mathrm{bp}}(s) := \frac{2\pi f_{\mathrm{h}} s}{s^2 + 2\pi(f_{\mathrm{h}} + f_{\mathrm{l}})s + 4\pi^2 f_{\mathrm{h}} f_{\mathrm{l}}}, \tag{4.2d}$$

where $\tau \in \mathbb{R}$ is the pitch actuator time constant, $\frac{dM_{\mathrm{flap}}}{d\theta} \in \mathbb{R}$ represents the change in blade flap-wise bending moment with respect to pitch angle, $f_{\mathrm{b}} \in \mathbb{R}$ is the natural frequency of the blade's first flap-wise mode and $D_{\mathrm{b}} \in \mathbb{R}$ is its aerodynamic damping ratio, while $f_{\mathrm{h}}, f_{\mathrm{l}} \in \mathbb{R}$ are the high and low corner frequencies, respectively, of the bandpass filter. The basic individual pitch control problem is shown in Fig. 4.2 and is based upon the following three-blade model:

$$\begin{bmatrix} \tilde{M}_1(s) \\ \tilde{M}_2(s) \\ \tilde{M}_3(s) \end{bmatrix} = \underbrace{\begin{bmatrix} G(s) & 0 & 0 \\ 0 & G(s) & 0 \\ 0 & 0 & G(s) \end{bmatrix}}_{P(s)} \begin{bmatrix} \tilde{\theta}_1(s) \\ \tilde{\theta}_2(s) \\ \tilde{\theta}_3(s) \end{bmatrix}. \tag{4.3}$$

In the interests of simplicity, the influence of the fixed turbine structural dynamics have not been included, but if required, these could be represented as additive disturbances on the bending moment channels. The next section introduces the three different IPCs employed in this study. These are shown in Fig. 4.3b, beginning first with the Coleman Transform-based controller.

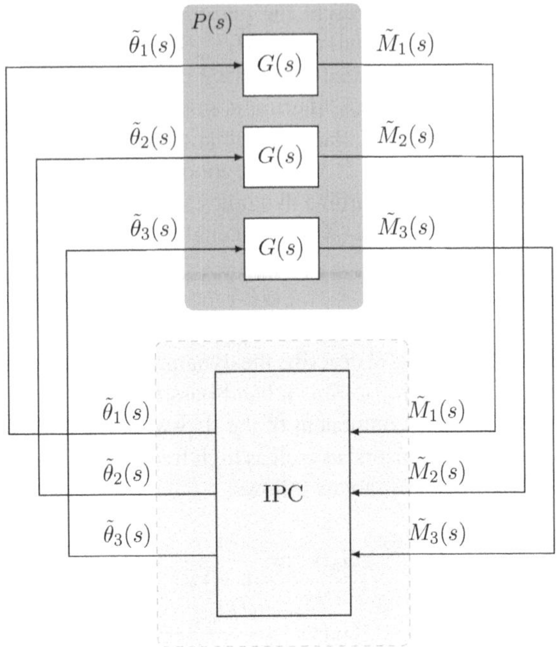

Fig. 4.2 Basic system architecture for IPC analysis and design

4.3.1 Coleman Transform-Based Control

The Coleman Transform-based controller is shown in Fig. 4.3a. As discussed in Sect. 4.2, many IPC studies have employed this form of IPC in order to attenuate unsteady loads upon the fixed turbine structure. The Coleman Transform $T_{\text{cm}}\,(\phi(t))$ is a time varying matrix that projects the rotational blade loads onto the stationary and orthogonal tilt and yaw axes of the turbine, according to the blade azimuth angle $\phi(t)$. For a three-bladed turbine in which $\phi(t)$ is defined as the angle of the first blade from the horizontal yaw axis, the Coleman Transform is defined as follows:

$$\begin{bmatrix} \tilde{M}_{\text{tilt}}(t) \\ \tilde{M}_{\text{yaw}}(t) \end{bmatrix} := \underbrace{\frac{2}{3} \begin{bmatrix} \sin\phi(t) & \sin\left(\phi(t) + \frac{2\pi}{3}\right) & \sin\left(\phi(t) + \frac{4\pi}{3}\right) \\ \cos\phi(t) & \cos\left(\phi(t) + \frac{2\pi}{3}\right) & \cos\left(\phi(t) + \frac{4\pi}{3}\right) \end{bmatrix}}_{T_{\text{cm}}\,(\phi(t))} \begin{bmatrix} \tilde{M}_1(t) \\ \tilde{M}_2(t) \\ \tilde{M}_3(t) \end{bmatrix}. \tag{4.4a}$$

The tilt and yaw referred flap-wise blade root bending moments, $\tilde{M}_{\text{tilt}}$ and $\tilde{M}_{\text{yaw}}$ are mapped via the Coleman controller $C_{\text{cm}} \in \mathcal{R}^{2\times 2}$ to tilt and yaw referred pitch signals $\tilde{\theta}_{\text{tilt}}$ and $\tilde{\theta}_{\text{yaw}}$, that in turn are projected back into the blade referred pitch signals via the inverse Coleman Transform, $T_{\text{cm}}^{\text{inv}}\,(\phi(t))$ accordingly:

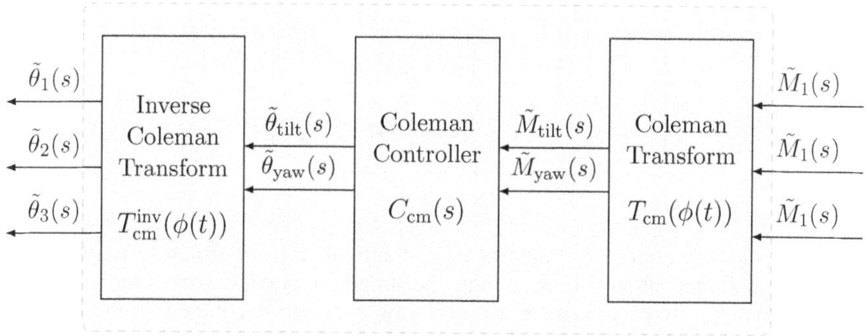

(A) Coleman Transform-based controller.

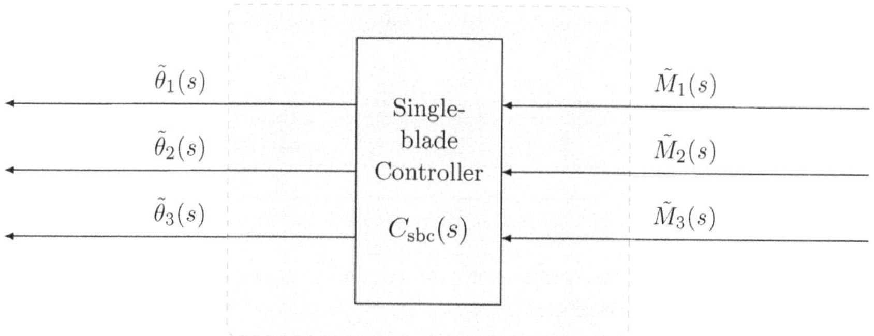

(B) Single-blade controller.

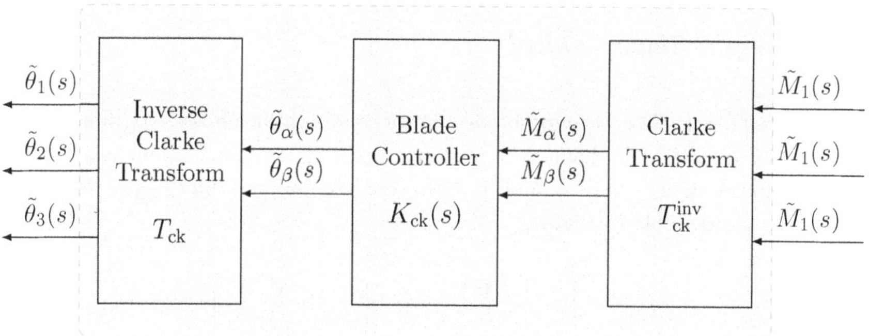

(C) Clarke Transform-based controller.

Fig. 4.3 Three different IPC architectures

$$\begin{bmatrix} \tilde{\theta}_1(t) \\ \tilde{\theta}_2(t) \\ \tilde{\theta}_3(t) \end{bmatrix} := \underbrace{\begin{bmatrix} \sin\phi(t) & \cos\phi(t) \\ \sin\left(\phi(t) + \frac{2\pi}{3}\right) & \cos\left(\phi(t) + \frac{2\pi}{3}\right) \\ \sin\left(\phi(t) + \frac{4\pi}{3}\right) & \cos\left(\phi(t) + \frac{4\pi}{3}\right) \end{bmatrix}}_{T_{\text{cm}}^{\text{inv}}(\phi(t))} \begin{bmatrix} \tilde{\theta}_{\text{tilt}}(t) \\ \tilde{\theta}_{\text{yaw}}(t) \end{bmatrix}. \tag{4.4b}$$

A basic Coleman controller consists of a diagonal transfer function matrix with equal proportional-integral terms along the diagonal. Such a controller implicitly assumes that the dynamics of the tilt and yaw axes are decoupled. However, this was shown not to be the case in Lu et al. (2015). By modelling the dynamics of the Coleman Transform and its inverse, Lu et al. (2015) showed how these operators modify the basic plant dynamics (4.3) to yield the Coleman-transformed plant $P_{\text{cm}} \in \mathcal{R}^{2\times 2}$:

$$\begin{bmatrix} \tilde{M}_{\text{tilt}}(s) \\ \tilde{M}_{\text{yaw}}(s) \end{bmatrix} = \underbrace{\begin{bmatrix} \frac{G(s+j\omega_0)+G(s-j\omega_0)}{2} & j\frac{G(s+j\omega_0)-G(s-j\omega_0)}{2} \\ -j\frac{G(s+j\omega_0)-G(s-j\omega_0)}{2} & \frac{G(s+j\omega_0)+G(s-j\omega_0)}{2} \end{bmatrix}}_{P_{\text{cm}}(s,\omega_0)} \begin{bmatrix} \tilde{\theta}_{\text{tilt}}(s) \\ \tilde{\theta}_{\text{yaw}}(s) \end{bmatrix}, \tag{4.5}$$

where $\omega_0 \in \mathbb{R}$ is the constant rated rotor speed, and from which the coupled nature of the tilt and yaw loops is evident.

4.3.2 Single-Blade Control

The simplest form of IPC is single-blade control, in which each blade is equipped with its own controller $K \in \mathcal{R}$ that acts in response to the local blade load measurements. Single-blade control is depicted in Fig. 4.3b, wherein the controller $C_{\text{sbc}} \in \mathcal{R}^{3\times 3}$ has the following decoupled structure:

$$\begin{bmatrix} \tilde{\theta}_1(s) \\ \tilde{\theta}_2(s) \\ \tilde{\theta}_3(s) \end{bmatrix} = \underbrace{\begin{bmatrix} K(s) & 0 & 0 \\ 0 & K(s) & 0 \\ 0 & 0 & K(s) \end{bmatrix}}_{C_{\text{sbc}}(K(s))} \begin{bmatrix} \tilde{M}_1(s) \\ \tilde{M}_2(s) \\ \tilde{M}_3(s) \end{bmatrix} \tag{4.6}$$

The blade controller K is designed to attenuate the blade loads at 1p, 2p and 4p frequencies. The benefits of this approach over those employing the Coleman Transform are that it can be realised as three, separate SISO controllers and also does not require a measurement of the rotor azimuth angle.

4.3.3 Clarke Transform-Based Control

Another IPC technique, based on blade load reductions, was recently introduced by Zhang et al. (2013) and employed the Clarke Transform to project the blade loads onto a pair of orthogonal axes that are stationary with respect to the turbine blades. Such a controller is shown in Fig. 4.3c, and consists of a diagonal blade controller $K_{\text{ck}} \in \mathcal{R}^{2\times 2}$ pre and post-compensated by the Clarke Transform $T_{\text{ck}} \in \mathbb{R}^{3\times 2}$ and its inverse $T_{\text{ck}}^{\text{inv}} \in \mathbb{R}^{2\times 3}$, as follows:

$$\begin{bmatrix} \tilde{\theta}_1(s) \\ \tilde{\theta}_2(s) \\ \tilde{\theta}_3(s) \end{bmatrix} := \underbrace{T_{\text{ck}}^{\text{inv}} K_{\text{ck}}(K(s)) T_{\text{ck}}}_{C_{\text{ck}}(K(s))} \begin{bmatrix} \tilde{M}_1(s) \\ \tilde{M}_2(s) \\ \tilde{M}_3(s) \end{bmatrix}, \tag{4.7a}$$

where:

$$T_{\text{ck}}^{\text{inv}} = \sqrt{\frac{2}{3}} \begin{bmatrix} 1 & 0 \\ -\frac{1}{2} & \frac{\sqrt{3}}{2} \\ -\frac{1}{2} & -\frac{\sqrt{3}}{2} \end{bmatrix}, \; K_{\text{ck}}(K(s)) = \begin{bmatrix} K(s) & 0 \\ 0 & K(s) \end{bmatrix}, \; T_{\text{ck}} = \sqrt{\frac{2}{3}} \begin{bmatrix} 1 & -\frac{1}{2} & -\frac{1}{2} \\ 0 & \frac{\sqrt{3}}{2} & -\frac{\sqrt{3}}{2} \end{bmatrix}. \tag{4.7b}$$

As with the single-blade controller, the blade controllers K in the Clarke controller C_{ck} are designed to minimise the loads at 1p, 2p and 4p frequencies, but do so upon the orthogonally projected blade load signals $\tilde{M}_\alpha(t)$ and $\tilde{M}_\beta(t)$, as opposed to $\tilde{M}_{1,2,3}(t)$. Similarly to the single-blade controller, the Clarke controller does not require a measurement of the blade azimuth angle and the control design amounts to the design of a single SISO blade controller. However, the Clarke controller achieves its load reductions using only two SISO controllers, suggesting a degree of redundancy exists in the single-blade controller (4.6).

4.4 Equivalence of Single-Blade, Coleman and Clarke Transform-Based Controllers

In this Section, for a given blade controller K, the equivalence between the blade load IPCs, $C_{\text{sbc}}(K)$ (4.6), $C_{\text{ck}}(K)$ (4.7) and a particular type of Coleman Transform-based controller C_{cm} is established. This leads to the main result of the chapter (Theorem 4.3) that proves that the performance of all three controllers is identical.

4.4.1 Equivalence Between Single-Blade and Coleman Transform-Based Control

The equivalence between single-blade control and Coleman Transform-based control is first established. This amounts to ascertaining the form that a single-blade controller takes when referred to tilt and yaw coordinates via the Coleman Transforms. The following lemma establishes this equivalence.

Lemma 4.1 *Assuming a constant rotor speed $\omega(t) = \omega_0$, Coleman Transforms (4.4) and a given blade controller K, a single-blade controller $C_{\text{sbc}}(K)$ (4.6) is equivalent to the Coleman Transform-based controller $C_{\text{cm}}(K, \omega_0)$, where:*

$$C_{\text{cm}}(K(s), \omega_0) := \begin{bmatrix} \dfrac{K(s+j\omega_0)+K(s-j\omega_0)}{2} & j\dfrac{K(s+j\omega_0)-K(s-j\omega_0)}{2} \\ -j\dfrac{K(s+j\omega_0)-K(s-j\omega_0)}{2} & \dfrac{K(s+j\omega_0)+K(s-j\omega_0)}{2} \end{bmatrix} \tag{4.8}$$

Proof The proof makes use of the following identities:

$$\mathcal{L}\left[u(t)\cos\phi(t)\right] = \mathcal{L}\left[u(t)\frac{e^{j\omega_0 t}+e^{-j\omega_0 t}}{2}\right] = \frac{1}{2}\left(U(s-j\omega_0)+U(s+j\omega_0)\right), \tag{4.9a}$$

$$\mathcal{L}\left[u(t)\sin\phi(t)\right] = \mathcal{L}\left[u(t)\frac{j\left(e^{-j\omega_0 t}-e^{j\omega_0 t}\right)}{2}\right] = \frac{j}{2}\left(U(s+j\omega_0)-U(s-j\omega_0)\right), \tag{4.9b}$$

where $u(t)$ is an arbitrary input signal, $U(s)$ is its Laplace transform and $\phi(t) = \omega_0 t$ is assumed. Inserting (4.9) into (4.4) yields:

$$\begin{bmatrix} \tilde{M}_1(s) \\ \tilde{M}_2(s) \\ \tilde{M}_3(s) \end{bmatrix} = C_-^T \begin{bmatrix} \tilde{M}_{\text{tilt}}(s-j\omega_0) \\ \tilde{M}_{\text{yaw}}(s-j\omega_0) \end{bmatrix} + C_+^T \begin{bmatrix} \tilde{M}_{\text{tilt}}(s+j\omega_0) \\ \tilde{M}_{\text{yaw}}(s+j\omega_0) \end{bmatrix}, \tag{4.10a}$$

$$\begin{bmatrix} \tilde{\theta}_{\text{tilt}}(s) \\ \tilde{\theta}_{\text{yaw}}(s) \end{bmatrix} = \frac{2}{3}C_- \begin{bmatrix} \tilde{\theta}_1(s-j\omega_0) \\ \tilde{\theta}_2(s-j\omega_0) \\ \tilde{\theta}_3(s-j\omega_0) \end{bmatrix} + \frac{2}{3}C_+ \begin{bmatrix} \tilde{\theta}_1(s+j\omega_0) \\ \tilde{\theta}_2(s+j\omega_0) \\ \tilde{\theta}_3(s+j\omega_0) \end{bmatrix}, \tag{4.10b}$$

where C_- and C_+ are defined as:

$$C_- := \frac{1}{2}\begin{bmatrix} 1 & -j \\ j & 1 \end{bmatrix} \begin{bmatrix} \sin(0) & \sin(\frac{2\pi}{3}) & \sin(\frac{4\pi}{3}) \\ \cos(0) & \cos(\frac{2\pi}{3}) & \cos(\frac{4\pi}{3}) \end{bmatrix}, \tag{4.10c}$$

$$C_+ := \frac{1}{2}\begin{bmatrix} 1 & j \\ -j & 1 \end{bmatrix} \begin{bmatrix} \sin(0) & \sin(\frac{2\pi}{3}) & \sin(\frac{4\pi}{3}) \\ \cos(0) & \cos(\frac{2\pi}{3}) & \cos(\frac{4\pi}{3}) \end{bmatrix}. \tag{4.10d}$$

Substituting (4.10) into (4.6) yields (4.8). □

It is interesting to note that the Coleman controller (4.8) possesses the same structure as the Coleman transformed plant (4.5), in much the same way as the single-blade controller (4.6) shares the diagonal structure of the turbine blade model (4.3). In view of this, the controller (4.8) will henceforth be termed a structured Coleman Transform-based controller.

4.4.2 Equivalence Between Structured Coleman Transform and Clarke Transform-Based Controllers

The projection from single-blade to tilt-yaw control via the Coleman Transforms yielded the structured Coleman Transform-based controller (4.8). However, the projection of (4.8) back to the rotating frame of reference does not yield the single-blade controller (4.6). Instead, it yields a Clarke Transform-based controller (4.7), according to the following lemma.

Lemma 4.2 *Assuming a constant rotor speed* $\omega(t) = \omega_0$, *Coleman Transforms (4.4) and a given blade controller* K, *the structured Coleman Transform-based controller* $C_{\mathrm{cm}}(K, \omega_0)$ (4.8) *is equivalent to* $C_{\mathrm{ck}}(K)$ *(4.7).*

Proof Referring to Fig. 4.3a and using the relationships (4.8) and (4.10), the derivation is as follows:

$$
\begin{aligned}
\begin{bmatrix} \tilde{\theta}_1(s) \\ \tilde{\theta}_2(s) \\ \tilde{\theta}_3(s) \end{bmatrix} &= C_-^T \begin{bmatrix} \tilde{\theta}_{\mathrm{tilt}}(s - j\omega_0) \\ \tilde{\theta}_{\mathrm{yaw}}(s - j\omega_0) \end{bmatrix} + C_+^T \begin{bmatrix} \tilde{\theta}_{\mathrm{tilt}}(s + j\omega_0) \\ \tilde{\theta}_{\mathrm{yaw}}(s + j\omega_0) \end{bmatrix}, \\
&= C_-^T C_{\mathrm{cm}}(s - j\omega_0) \begin{bmatrix} \tilde{M}_{\mathrm{tilt}}(s - j\omega_0) \\ \tilde{M}_{\mathrm{yaw}}(s - j\omega_0) \end{bmatrix} + C_+^T C_{\mathrm{cm}}(s + j\omega_0) \begin{bmatrix} \tilde{M}_{\mathrm{tilt}}(s + j\omega_0) \\ \tilde{M}_{\mathrm{yaw}}(s + j\omega_0) \end{bmatrix}, \\
&= \frac{2}{3} \left[C_-^T C_{\mathrm{cm}}(s - j\omega_0) \left(C_- \begin{bmatrix} \tilde{M}_1(s - 2j\omega_0) \\ \tilde{M}_2(s - 2j\omega_0) \\ \tilde{M}_3(s - 2j\omega_0) \end{bmatrix} + C_+ \begin{bmatrix} \tilde{M}_1(s) \\ \tilde{M}_2(s) \\ \tilde{M}_3(s) \end{bmatrix} \right) + \dots \right. \\
&\quad \left. \dots + C_+^T C_{\mathrm{cm}}(s + j\omega_0) \left(C_- \begin{bmatrix} \tilde{M}_1(s) \\ \tilde{M}_2(s) \\ \tilde{M}_3(s) \end{bmatrix} + C_+ \begin{bmatrix} \tilde{M}_1(s + 2j\omega_0) \\ \tilde{M}_2(s + 2j\omega_0) \\ \tilde{M}_3(s + 2j\omega_0) \end{bmatrix} \right) \right], \\
&= \begin{bmatrix} \frac{2}{3}K(s) & -\frac{1}{3}K(s) & -\frac{1}{3}K(s) \\ -\frac{1}{3}K(s) & \frac{2}{3}K(s) & -\frac{1}{3}K(s) \\ -\frac{1}{3}K(s) & -\frac{1}{3}K(s) & \frac{2}{3}K(s) \end{bmatrix} \begin{bmatrix} \tilde{M}_1(s) \\ \tilde{M}_2(s) \\ \tilde{M}_3(s) \end{bmatrix}, \\
&= C_{\mathrm{ck}}(s) \begin{bmatrix} \tilde{M}_1(s) \\ \tilde{M}_2(s) \\ \tilde{M}_3(s) \end{bmatrix}.
\end{aligned}
$$

□

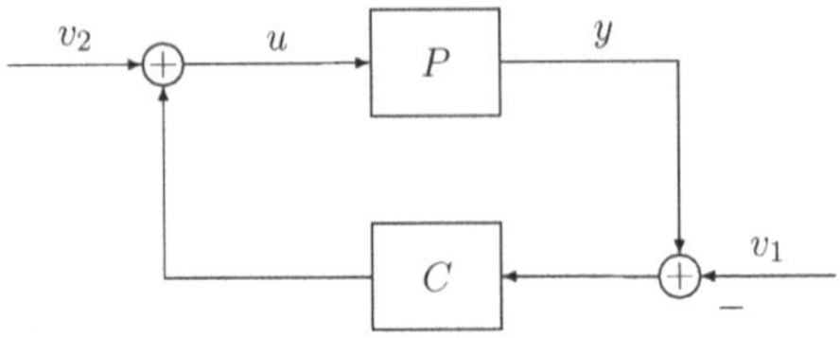

Fig. 4.4 Standard feedback interconnection between plant P and controller C. The signals u and y denote the plant input and measured output, respectively, whilst v_1 and v_2 represent exogenous disturbances

At this point the separate relationships have been established between a structured Coleman Transform-based controller, and single-blade and Clarke Transform-based controllers, respectively. The next section establishes the extent to which these three types of IPC behave in a similar fashion, as quantified by the robust stability margin.

4.4.3 Performance Equivalence of C_{sbc}, C_{ck} and C_{cm}

To examine performance equivalence of the IPC controllers, the concept of robust stability margin is introduced.

4.4.3.1 Robust Stability Margin

The standard feedback interconnection $[P, C]$ of plant $P \in \mathcal{R}$ and controller $C \in \mathcal{R}$ is shown in Fig. 4.4, from which the following closed-loop system is defined:

$$\begin{bmatrix} y \\ u \end{bmatrix} = \underbrace{\begin{bmatrix} P \\ I \end{bmatrix} (I - CP)^{-1} \begin{bmatrix} -C & I \end{bmatrix}}_{H(P,C)} \begin{bmatrix} v_1 \\ v_2 \end{bmatrix}, \tag{4.11}$$

where $H(P, C) \in \mathcal{R}$ provided $[P, C]$ is well posed, and I is an identity matrix of compatible dimension. The robust stability margin $b(P, C) \in \mathbb{R}$ of $[P, C]$ is defined as follows:

$$b(P, C) := \begin{cases} \|H(P, C)\|_\infty^{-1} & \text{if } H(P, C) \in \mathcal{RH}_\infty \\ 0 & \text{otherwise.} \end{cases} \tag{4.12}$$

4.4.3.2 Performance Equivalence

The main result of this chapter is as follows:

Theorem 4.3 *For a given blade model G (4.2) assume the turbine model $P(G)$ (4.3), and for a given fixed rotor speed ω_0 and blade controller K, form the IPC controllers*

$C_{\text{sbc}}(K)$, $C_{\text{ck}}(K)$ and $C_{\text{cm}}(K, \omega_0)$ according to (4.6), *(4.7) and* (4.8), *respectively. Then the robust stability margin for each IPC is the same. Specifically,*

$$b(GK, 1) = b(PC_{\text{sbc}}, I) = b(PC_{\text{ck}}, I) = b(PC_{\text{cm}}, I). \tag{4.13}$$

Proof The proof is based on the derivation and comparison of the $\mathcal{H}_\infty$-norms of the shaped systems $H(PC_{\text{sbc}}, I)$, $H(PC_{\text{ck}}, I)$ and $H(PC_{\text{cm}}, I)$. Proceeding with the former we obtain:

$$\|H(PC_{\text{sbc}}, I)\|_\infty := \left\| \begin{bmatrix} C_{\text{sbc}}P \\ I \end{bmatrix} (I - C_{\text{sbc}}P)^{-1} \begin{bmatrix} -I & I \end{bmatrix} \right\|_\infty$$
$$= \left\| \begin{bmatrix} T & 0 & 0 & -T & 0 & 0 \\ 0 & T & 0 & 0 & -T & 0 \\ 0 & 0 & T & 0 & 0 & -T \\ S & 0 & 0 & -S & 0 & 0 \\ 0 & S & 0 & 0 & -S & 0 \\ 0 & 0 & S & 0 & 0 & -S \end{bmatrix} \right\|_\infty,$$

where $S(j\omega) := 1/(GK - 1)(j\omega)$ and $T := GK/(GK - 1)(j\omega)$ denote the sensitivity and complementary sensitivity functions, respectively. We are concerned with the spectrum of the following operator:

$$H(PC_{\text{sbc}}, I)^* H(PC_{\text{sbc}}, I) = \begin{bmatrix} X_{11} & X_{12} \\ X_{21} & X_{22} \end{bmatrix}, \tag{4.14}$$

where:

$$X_{11} = -X_{12} = -X_{21} = X_{22} = \begin{bmatrix} S^*S + T^*T & 0 & 0 \\ 0 & S^*S + T^*T & 0 \\ 0 & 0 & S^*S + T^*T \end{bmatrix}$$

Next, noting that all four sub-matrices commute, the characteristic polynomial of (4.14) can be expressed as follows (Silvester, 2000):

$$\det\left(\lambda I - H(PC_{\text{sbc}}, I)^* H(PC_{\text{sbc}}, I)\right) = (\lambda I - X_{11})(\lambda I - X_{22}) - X_{12}X_{21}$$
$$= \lambda^3(\lambda - 2(S^*S + T^*T))^3.$$

The $\mathcal{H}_\infty$ norm of $H(PC_{\text{sbc}}, I)$ is therefore:

$$\|H(PC_{\text{sbc}}, I)\|_\infty = \sup_\omega \sqrt{2\,(S^*S + T^*T)} = \|H(GK, 1)\|_\infty . \tag{4.15}$$

Turning attention to $H(PC_{\text{ck}}, I)$, we begin by taking the singular value decomposition of C_{ck}:

$$C_{\text{ck}}(j\omega) = \underbrace{\begin{bmatrix} -\sqrt{2/3} & 0 & 1/\sqrt{3} \\ 1/\sqrt{6} & -1/\sqrt{2} & 1/\sqrt{3} \\ 1/\sqrt{6} & 1/\sqrt{2} & 1/\sqrt{3} \end{bmatrix}}_{U_{\text{ck}}} \underbrace{\begin{bmatrix} K(j\omega) & 0 & 0 \\ 0 & K(j\omega) & 0 \\ 0 & 0 & 0 \end{bmatrix}}_{\tilde{C}_{\text{ck}}(j\omega)} \underbrace{\begin{bmatrix} -\sqrt{2/3} & 1/\sqrt{6} & 1/\sqrt{6} \\ 0 & -1/\sqrt{2} & 1/\sqrt{2} \\ -1/\sqrt{6} & -1/\sqrt{6} & -1/\sqrt{6} \end{bmatrix}}_{V^*_{\text{ck}}}$$

Inserting this into $\|H(PC_{\text{ck}}, I)\|_\infty$ yields:

$$\begin{aligned} \|H(PC_{\text{ck}}, I)\|_\infty &:= \left\| \begin{bmatrix} C_{\text{ck}}P \\ I \end{bmatrix} (I - C_{\text{ck}}P)^{-1} \begin{bmatrix} -I & I \end{bmatrix} \right\|_\infty, \\ &= \left\| \begin{bmatrix} \tilde{C}_{\text{ck}}P \\ U^*_{\text{ck}}V_{\text{ck}} \end{bmatrix} (U^*_{\text{ck}}V_{\text{ck}} - \tilde{C}_{\text{ck}}P)^{-1} \begin{bmatrix} -U^*_{\text{ck}}V_{\text{ck}} & U^*_{\text{ck}}V_{\text{ck}} \end{bmatrix} \right\|_\infty, \\ &= \left\| \tilde{H}(PC_{\text{ck}}, I) \right\|_\infty, \end{aligned}$$

where:

$$\tilde{H}(PC_{\text{ck}}, I) := \begin{bmatrix} -T & 0 & 0 & -T & 0 & 0 \\ 0 & -T & 0 & 0 & -T & 0 \\ 0 & 0 & 0 & 0 & 0 & 0 \\ -S & 0 & 0 & -S & 0 & 0 \\ 0 & -S & 0 & 0 & -S & 0 \\ 0 & 0 & -1 & 0 & 0 & -1 \end{bmatrix}.$$

It can be shown that the characteristic polynomial of $\tilde{H}(PC_{\text{ck}}, I)^*\tilde{H}(PC_{\text{ck}}, I)$ is given by:

$$\det(\lambda I - \tilde{H}(PC_{\text{ck}}, I)^*\tilde{H}(PC_{\text{ck}}, I)) = \lambda^3(\lambda - 2)(\lambda - 2(S^*S + T^*T))^2.$$

The relative degree of G ensures $\sup_\omega(S^*S + T^*T) \geq 1$, hence:

$$\|H(PC_{\text{ck}}, I)\|_\infty = \sup_\omega \sqrt{2\,(S^*S + T^*T)}. \tag{4.16}$$

With respect to $H(P_{\text{cm}}C_{\text{cm}}, I)$, the singular value decomposition of P_{cm} is as follows:

$$P_{\text{cm}}(j\omega, \omega_0) = \underbrace{\begin{bmatrix} \frac{j}{\sqrt{2}} & \frac{-j}{\sqrt{2}} \\ \frac{1}{\sqrt{2}} & \frac{1}{\sqrt{2}} \end{bmatrix}}_{U_{\text{cm}}} \underbrace{\begin{bmatrix} G(j(\omega - \omega_0)) & 0 \\ 0 & G(j(\omega + \omega_0)) \end{bmatrix}}_{\tilde{P}_{\text{cm}}(j\omega, \omega_0)} \underbrace{\begin{bmatrix} \frac{-j}{\sqrt{2}} & \frac{1}{\sqrt{2}} \\ \frac{j}{\sqrt{2}} & \frac{1}{\sqrt{2}} \end{bmatrix}}_{U^*_{\text{cm}}}$$

Similarly, $C_{\text{cm}} = U_{\text{cm}}\tilde{C}_{\text{cm}}U^*_{\text{cm}}$, where:

$$\tilde{C}_{\mathrm{cm}}(j\omega, \omega_0) := \begin{bmatrix} K(j(\omega-\omega_0)) & 0 \\ 0 & K(j(\omega+\omega_0)) \end{bmatrix}$$

Inserting these into $\|H(P_{\mathrm{cm}}C_{\mathrm{cm}}, I)\|_\infty$ yields:

$$\begin{aligned} \|H(PC_{\mathrm{cm}}, I)\|_\infty &:= \left\| \begin{bmatrix} C_{\mathrm{cm}}P_{\mathrm{cm}} \\ I \end{bmatrix} (I - C_{\mathrm{cm}}P_{\mathrm{cm}})^{-1}[-I\ I] \right\|_\infty, \\ &= \left\| \begin{bmatrix} \tilde{C}_{\mathrm{cm}}\tilde{P}_{\mathrm{cm}} \\ I \end{bmatrix} (I - \tilde{C}_{\mathrm{cm}}\tilde{P}_{\mathrm{cm}})^{-1}[-I\ I] \right\|_\infty, \\ &= \left\| \tilde{H}(PC_{\mathrm{cm}}, I) \right\|_\infty, \end{aligned}$$

in which:

$$\tilde{H}(PC_{\mathrm{cm}}, I) := \begin{bmatrix} T_- & 0 & -T_- & 0 \\ 0 & T_+ & 0 & -T_+ \\ S_- & 0 & -S_- & 0 \\ 0 & S_+ & 0 & -S_+ \end{bmatrix},$$

where $S_-(j\omega, \omega_0) := 1/(GK-1)(j(\omega-\omega_0))$ and $S_+(j\omega, \omega_0) := 1/(GK-1)(j(\omega+\omega_0))$ are the frequency shifted sensitivity functions, and $T_-(j\omega, \omega_0) := GK/(GK-1)(j(\omega-\omega_0))$ and $T_+(j\omega, \omega_0) := GK/(GK-1)(j(\omega+\omega_0))$ are the shifted complimentary sensitivity functions. It can be shown that the characteristic polynomial of $\tilde{H}(P_{\mathrm{cm}}C_{\mathrm{cm}}, I)^*\tilde{H}(P_{\mathrm{cm}}C_{\mathrm{cm}}, I)$ is given by:

$$\det(\lambda I - \tilde{H}(P_{\mathrm{cm}}C_{\mathrm{cm}}, I)^*\tilde{H}(P_{\mathrm{cm}}C_{\mathrm{cm}}, I)) = \lambda^2(\lambda - 2(S_-^*S_- + T_-^*T_-))(\lambda - 2(S_+^*S_+ + T_+^*T_+)).$$

The $\mathcal{H}_\infty$ norm of $H(P_{\mathrm{cm}}C_{\mathrm{cm}}, I)$ is thus given by:

$$\begin{aligned} \|H(P_{\mathrm{cm}}C_{\mathrm{cm}}, I)\|_\infty &= \sup_\omega \sqrt{2\left(S_-^*S_- + T_-^*T_-\right)} = \sup_\omega \sqrt{2\left(S_+^*S_+ + T_+^*T_+\right)} \\ &= \sup_\omega \sqrt{2\left(S^*S + T^*T\right)}. \end{aligned} \tag{4.17}$$

□

This suggests that the three different IPC strategies studied in this chapter behave in exactly the same fashion. This is indeed the case, as shown in the following section.

4.5 Numerical Results and Discussion

The objective of this section is to demonstrate the performance equivalence of the various IPCs by performing closed-loop simulations of each controller upon upon a high-fidelity wind turbine model. The turbine model employed for this purpose is the NREL 5 MW baseline turbine (Jonkman, Butterfield, Musial, & Scott, 2009),

Table 4.1 Turbine simulation parameters

Rating	5 MW
Rotor Orientation	Upwind
Rotor diameter	126 m
Hub height	90 m
Rated rotor speed	12.1 rpm (≈0.2 Hz)

with the key parameters listed in Table 4.1. Note that this model is of much greater complexity than the model employed for IPC design (4.3), and includes flap-wise and edge-wise blade modes, in addition to tower and shaft dynamics. The simulations were performed at an above-rated mean wind speed of 18 ms^{-1} and were run for sufficient duration to obtain convergence in the load spectra of the various key rotating and non-rotating turbine components.

4.5.1 *Blade Controller* $K(s)$

The three IPCs studied in this chapter, (4.6), (4.7) and (4.8) are each a function of the underlying blade controller K. In turn, the design of K is based upon the basic blade model G (4.2a). The resultant blade controller K is the same as the baseline IPC controller (2.56) in Sect. 2.8, where the controller (2.56) attenuates blade loads specifically at the 1p, 2p and 4p frequencies (0.2, 0.4 and 0.8 Hz), respectively, and yielded a robust stability margin $b(GK, 1) = 0.39$. Based on this controller, the IPCs (4.6), (4.7) and (4.8) were generated and tested in simulation, as shown next.

4.5.2 *IPC Simulation Results upon the NREL 5 MW Turbine*

Closed-loop simulations were performed upon each IPC and results were obtained to compare the load reductions on both the blades as well as the fixed turbine structures. Figure 4.6a shows the power spectrum of the flap-wise blade bending moment upon a particular blade, whilst Fig. 4.6b, c display the power spectra of the main bearing tilt and yaw bending moments. With respect to the blade loads (Fig. 4.6a), the performance of the separate IPCs are almost identical and display clear load reductions around the 1p and 2p frequencies, as compared to the uncontrolled turbine. In addition, there are further slight reductions at the 4p frequency. This is to be expected given the designed loop-shape of GK, as shown in Fig. 4.5. The load reductions at these frequencies translate to reductions at 0p and 3p frequencies in the fixed turbine structures as is evident from Fig. 4.6b, c, where again, the performance of the separate IPCs are almost indistinguishable. Given the performance similarities, it is no surprise that the pitch activity from each IPC is almost identical, as shown in Fig. 4.6d.

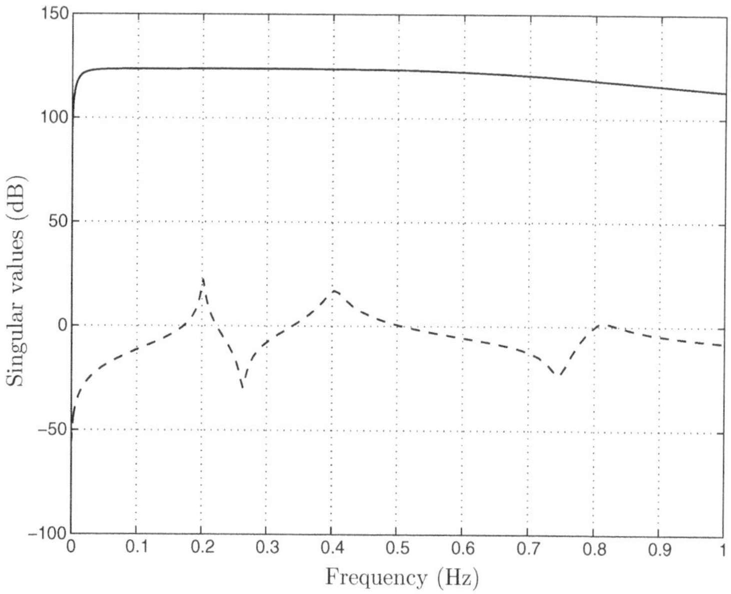

Fig. 4.5 Maximum singular value plots of the wind turbine model $\bar{\sigma}(G(s))$ (–) and compensated system $\bar{\sigma}(GK(s))$ (- -)

There is an important detail to note at this point. Close inspection of the results displayed in Fig. 4.6 reveals that although the performance of the three IPCs is almost identical, there nevertheless exist some small differences, particularly between the structured Coleman Transform-based controller and its counterparts. This is at odds with Theorem 4.3, which suggests that there should be no performance difference. The reasons for this are explained next.

4.5.3 Discussion

The slight discrepancies in IPC performance arise from an assumption of the turbine operating with a constant rotor speed. In practice, this is difficult to achieve owing to the limitations of the CPC, in addition to the coupling between CPC and IPC through the tower dynamics (Selvam et al. 2009). This challenge to maintaining fixed rotor speed can clearly be seen in Fig. 4.6d for the case without IPC, where changes in rotor speed are causing the CPC to continuously adjust the blade pitch angle. The structured Coleman Transform-based controller (4.8) is designed based upon an assumption of fixed rotor speed, and so perturbations to the rotor speed will inevitably result in deterioration in controller performance, although this is likely to be very small. To demonstrate this is indeed the case, the simulations of Sect. 4.5.2 were repeated, but in the absence of tower dynamics. This cancels the fore-aft motion

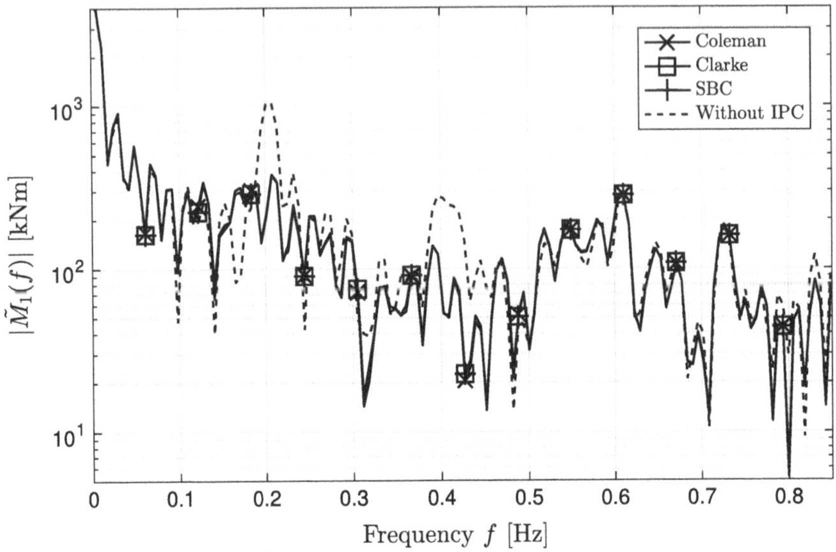

(A) Power spectrum of the flap-wise blade bending moment of blade 1.

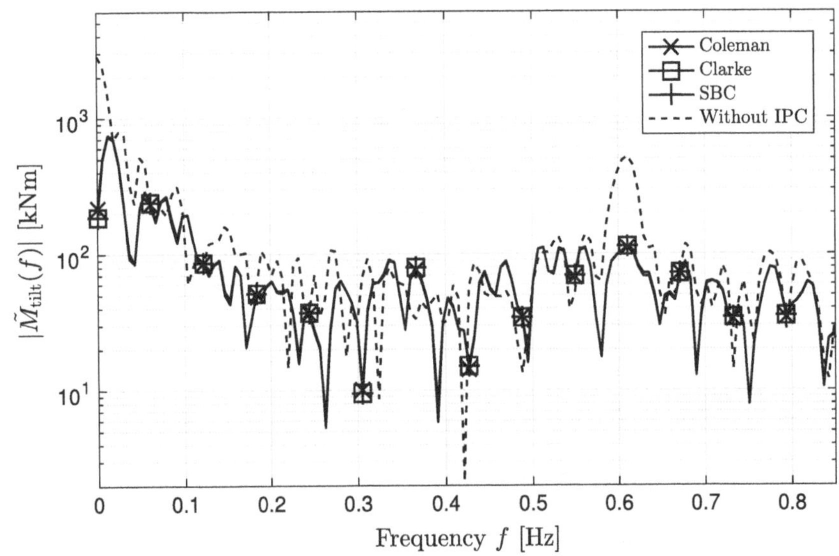

(B) Power spectrum of the main bearing tilt bending moment.

Fig. 4.6 Simulation results upon the NREL 5 MW turbine, showing the performance similarities between the various IPCs

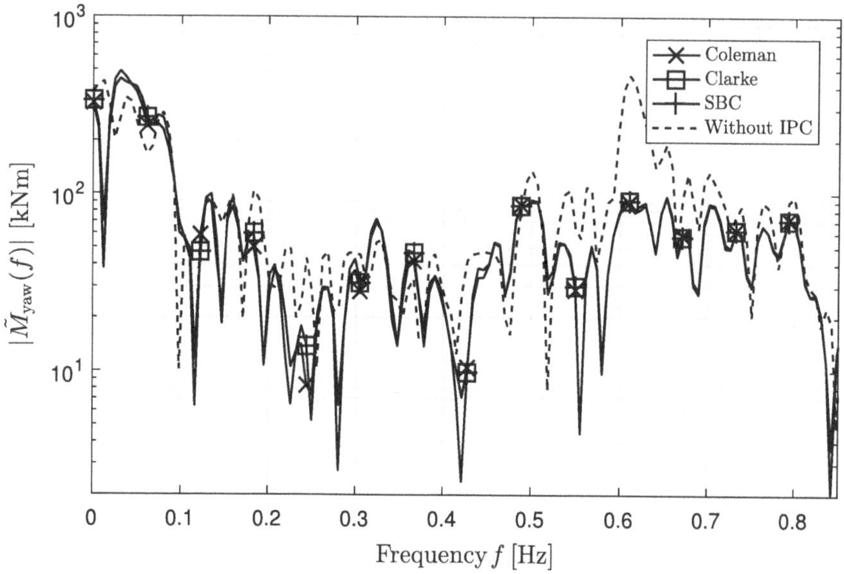

(c) Power spectrum of the main bearing yaw bending moment. Similar results are observed as in Fig 4.6b.

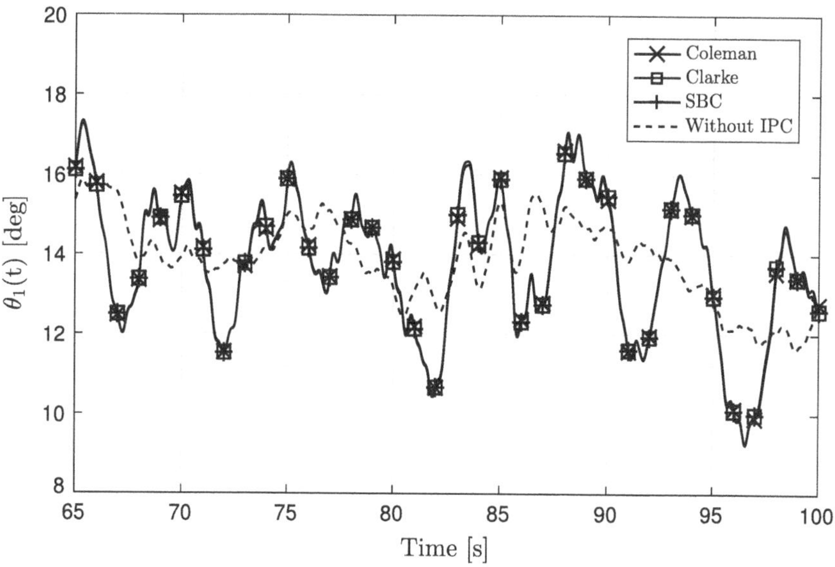

(d) Time history of the blade-pitch angle of blade 1.

Fig. 4.6 (continued)

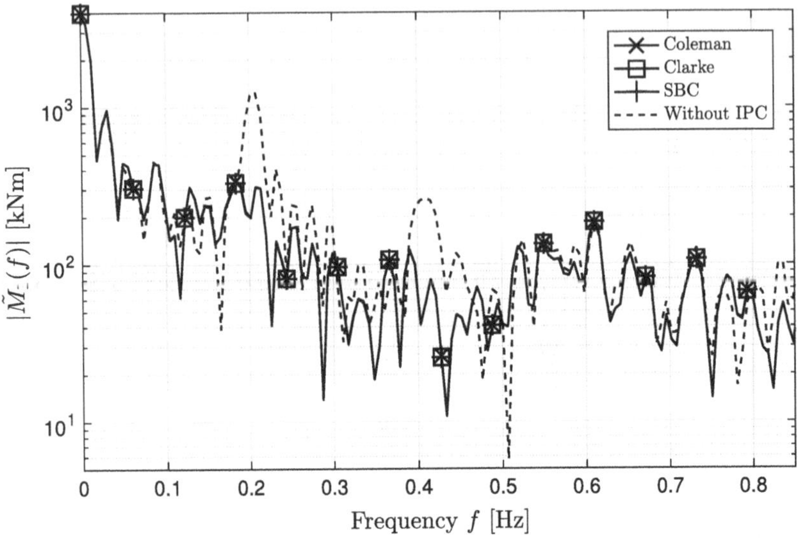

(A) Power spectrum of the flap-wise blade root bending moment of blade 1, with fixed rotor speed.

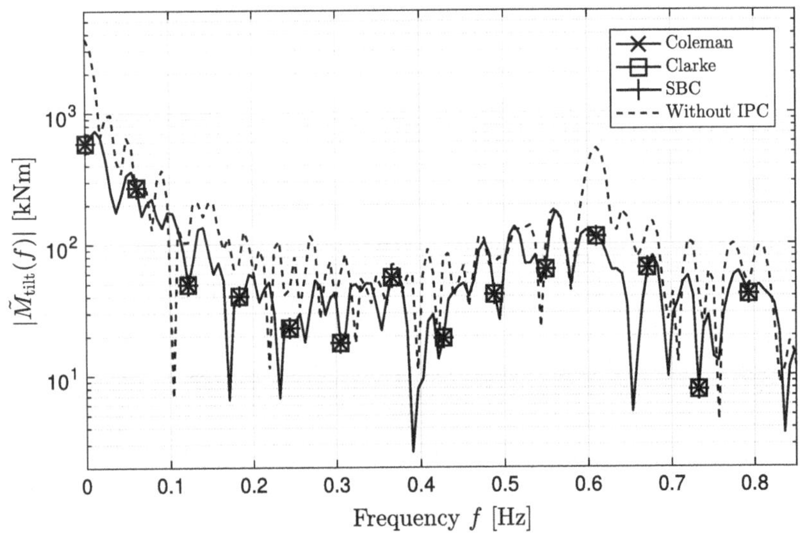

(B) Power spectrum of the main bearing tilt bending moment with fixed rotor speed.

Fig. 4.7 Simulation results upon the NREL 5 MW turbine with fixed rotor speed, showing indistinguishable performance between the various IPCs

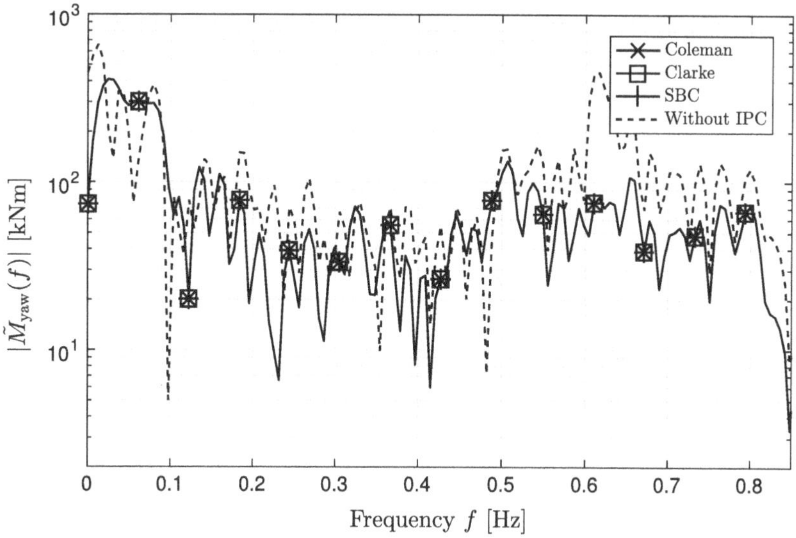

(c) Power spectrum of the main bearing yaw bending moment with fixed rotor speed.

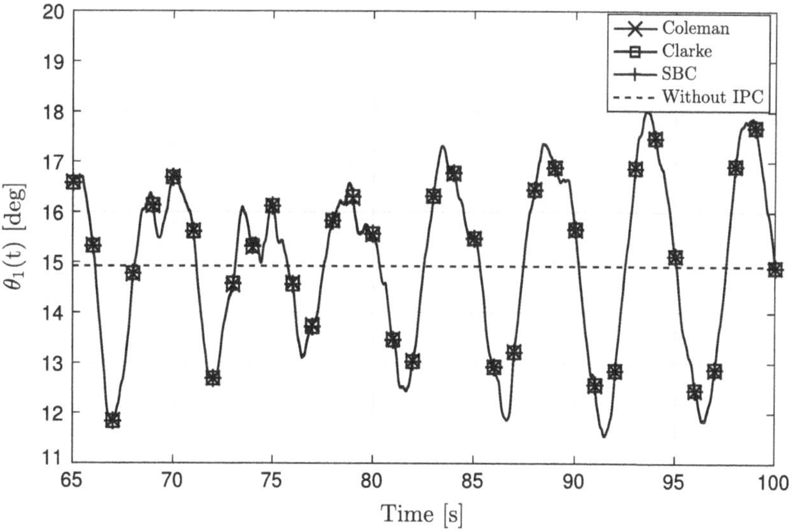

(d) Time history of the blade-pitch angle of blade 1 with fixed rotor speed.

Fig. 4.7 (continued)

of the turbine and thus eliminates a major source of disturbance to the collective-pitch loop that regulates the rotor speed. With this in mind, Fig. 4.7 displays the load spectra spectra and pitch activity, from which it is clear that the performance of the various IPCs is indistinguishable.

Given the essentially identical performance from the various IPCs, the question of 'which is best' is not straightforward to answer, and may rest with issues of implementation and load design priorities. For instance, the implementation of single-blade control is arguably the simplest; essentially amounting to the installation of three identical SISO control systems. On the other hand, the implementation of Coleman and Clarke Transform-based controllers is slightly more involved, with both being MIMO and the Coleman controller in particular requiring a measurement of the rotor azimuth angle. However, if load reductions on the fixed turbine structure are a priority, then the natural environment in which to design such a controller is in the tilt and yaw frame of reference, motivating the design of a structured Coleman Transform-based controller. Of course, this could then be referred back to the rotating frame of reference for implementation as either a single-blade or Clarke Transform-based controller, via the relationships established in Lemmas 4.1 and 4.2.

4.6 Summary

This chapter established the links between three different IPC techniques; those based on the Clarke and Coleman Transforms and single-blade control. The equivalence between single-blade and a structured Coleman Transform-based controller was established, as was the equivalence between the latter and Clarke-Transform-based control. Under an assumption of fixed rotor speed, analytical and numerical results were presented that showed no performance difference between these IPCs, as quantified by the robust stability margin. Choice of IPC thus largely rests with preference of design and implementation.

The coming chapter will look to accommodate the influence of tower motion in the design of IPCs, with a view towards removing the need for measurements of tower fore-aft motion. It is surmised that particular IPC architectures may lend themselves more readily to achieving this, and so may yet influence the issue of 'best' choice of IPC.

References

Barlas, T., & van Kuik, G. (2010). Review of state of the art in smart rotor control research for wind turbines. *Progress in Aerospace Sciences*, *46*(1), 1–27.

Bossanyi, E. A. (2003a). Individual Blade Pitch Control for Load Reduction. *Wind Energy*, *6*(2), 119–128.

Bossanyi, E. A. (2005). Further load reductions with individual pitch control. *Wind Energy*, *8*(4), 481–485.

Bossanyi, E. A., & Wright, A. (2009). Field testing of individual pitch control on the NREL CART-2 wind turbine. In *European Wind Energy Conference*.

Coleman, R. P., & Feingold, A. M. (1957). Theory of self-excited mechanical oscillations of helicopter rotors with hinged blades. *National Advisory Committee for Aeronautics*, Report 1351.

Engels, W. P., Subhani, S., Zafar, H., & Savenije, F. (2014). Extending wind turbine operational conditions; a comparison of set point adaptation and LQG individual pitch control for highly turbulent wind. In *The science of making torque from wind* (Vol. 524, p. 012058).

Geyler, M., & Caselitz, P. (2008). Robust multivariable pitch control design for load reduction on large wind turbines. *Journal of Solar Energy Engineering*, *130*(3), 031014.

Han, Y., & Leithead, W. E. (2014). Combined wind turbine fatigue and ultimate load reduction by individual blade control. In *Proceedings of the Science of Making Torque from Wind* (Vol. 524, p. 012062).

Jonkman, J., Butterfield, S., Musial, W., & Scott, G. (2009). *Definition of a 5-MW reference wind turbine for offshore system development* (Technical Report). Golden, CO: National Renewable Energy Laboratory (NREL).

Lackner, M. A., & van Kuik, G. (2010). A comparison of smart rotor control approaches using trailing edge aps and individual pitch control. *Wind Energy*, *13*(2–3), 117–134.

Leithead, W., Neilson, V., & Dominguez, S. (2009). Alleviation of unbalanced rotor loads by single blade controllers. In *European Wind Energy Conference*.

Leithead, W., Neilson, V., Dominguez, S., & Dutka, A. (2009). A novel approach to structural load control using intelligent actuators. In *2009 17th Mediterranean Conference on Control and Automation* (pp. 1257–1262). IEEE.

Lu, Q., Bowyer, R., & Jones, B. (2015). Analysis and design of Coleman transformbased individual pitch controllers for wind-turbine load reduction. *Wind Energy*, *18*(8), 1451–1468.

Muljadi, E. (2001). Pitch-controlled variable-speed wind turbine generation. *IEEE Transactions on Industry Applications*, *37*(1), 240–246.

Pao, L., & Johnson, K. (2009). A tutorial on the dynamics and control of wind turbines and wind farms. In *Proceedings of ACC*.

Plumley, C., Leithead, W., Jamieson, P., Bossanyi, E., & Graham, M. (2014). Comparison of individual pitch and smart rotor control strategies for load reduction. In *Proceedings of the Science of Making Torque from Wind* (Vol. 524, p. 012054).

Selvam, K., Kanev, S., van Wingerden, J. W., van Engelen, T., & Verhaegen, M. (2009). Feedback-feedforward individual pitch control for wind turbine load reduction. *International Journal of Robust and Nonlinear Control*, *19*(1), 72–91.

Silvester, J. R. (2000). Determinants of block matrices. *The Mathematical Gazette*, *84*(501), 460.

Stol, K., Moll, H., Bir, G., & Namik, H. (2009). A comparison of multi-blade coordinate transformation and direct periodic techniques for wind turbine control design. In *Proceedings of 47th AIAA/ASME* (pp. 1–12).

van Engelen, T. G. (2006). Design model and load reduction assessment for multirotational mode individual pitch control (higher harmonics control). In *Proceedings of European Wind Energy Conference*.

van Engelen, T. G., & van der Hooft, E. L. (2005). *Individual Pitch Control Inventory* (Technical Report). ECN.

Vas, P. (1992). *Electrical machines and drives: A space-vector theory approach*. Oxford: Oxford University Press.

Zhang, Y., Cheng, M., & Chen, Z. (2013). Proportional resonant individual pitch control for mitigation of wind turbines loads. *IET Renewable Power Generation*, *7*(3), 191–200.

Chapter 5
Estimation and Control Design for Tower Motions

5.1 Overview and Contribution

Wind turbine blade pitch adjustment can be used to alleviate excessive tower loads caused by unsteady aerodynamic loads. At present, tower damping control strategies assume sensory input from an additional tower-top accelerometer, leading to extra costs associated with installation. The main result of this chapter is to show that this sensor is redundant in the sense that tower motion can be estimated solely from existing blade load sensors. This is possible owing to the dynamic coupling between the blades and towers, but is challenging since the blade loads occur in a rotating frame of reference with respect to the fixed tower, resulting in a time-varying system. However, this chapter shows how a judicious choice of coordinate transformation simplifies the turbine structural system description to one that is time-invariant, greatly simplifying the observability analysis and subsequent observer design. As a further result, this chapter then shows how existing blade-pitch control architectures can be modified to accommodate and exploit the tower motion estimates. Closed-loop simulation results are presented that show the observer-based tower damping controller achieving load reductions of up to 87% at the tower resonant frequency for a modest increase of 9% in the standard deviation of the blade pitch rates.

The remainder of this chapter is structured as follows. Section 5.2 presents an introduction and motivation of the chapter. Section 5.3 provides a background of the turbine blade and tower structure modelling. In Sect. 5.4, analysis of the observability of the time-periodic modelled system is presented. In Sect. 5.5, the Coleman Transform is used to derive a linear time-invariant (LTI) turbine model, whose observability is then established. The design of a subsequent tower-top motion estimator and tower damping controller is described in Sect. 5.6. In Sect. 5.7, the performance of the proposed estimator and controller are demonstrated in simulation upon a high-fidelity and nonlinear wind turbine model. Finally, Sect. 5.8 concludes the chapter with a summary and an overview of future work.

W. H. (Alan) Lio, *Blade-Pitch Control for Wind Turbine Load Reductions*, Springer Theses, https://doi.org/10.1007/978-3-319-75532-8_5

5.2 Introduction

Large wind turbines experience uneven and intermittent aerodynamic loads from the wind and such loads inevitably contribute to fatigue damage upon the turbine structures. In order to manage the competing demands of power capture and load mitigation, most modern turbines employ a combination of control systems based upon blade pitch actuation. Primary amongst these is the use of collective pitch control (CPC) (Pao & Johnson, 2009), whereby the pitch angle of each blade is adjusted by an equal amount in order to regulate the rotor speed in above rated conditions. In addition, individual pitch control (IPC) and tower damping control can be used to specifically attenuate unsteady loads that play no part in power generation. The IPC provides additional pitch demand signals to each blade in order to balance the loads across the rotor plane, typically in response to measurements of the flap-wise blade root bending moments (Bossanyi, 2003a; Dunne, Pao, Wright, Jonkman, & Kelley, 2011; Laks, Pao, Wright, Kelley, & Jonkman, 2011; Leithead, Neilson, & Dominguez, 2009; Selvam, Kanev, van Wingerden, van Engelen, & Verhaegen, 2009), whilst tower damping control provides a further adjustment to the collective blade pitch angle in order to reduce excessive tower vibrations, in response to the tower fore-aft velocity measurements (Bossanyi, 2000, 2003b; Evans, Cannon, & Kouvaritakis, 2015; Larsen & Hanson, 2007; Leithead, Dominguez, & Spruce, 2004; Spruce & Turner, 2013). Typically, and for reasons of simplicity of implementation favoured by the industry, IPCs and tower damping controllers are designed separately from the CPC, and carefully in order to avoid cross-excitation (Leithead, Neilson, Dominguez, & Dutka, 2009; Lio, Jones, Lu, & Rossiter, 2015; Lu, Bowyer, & Jones, 2015; van Engelen, 2006).

At present, most tower damping control strategies assume a direct measurement of tower motion, typically from a nacelle-mounted accelerometer (e.g. Burton, Jenkins, Sharpe, & Bossanyi, 2011; Leithead, Neilson, & Dominguez, 2009). However, the turbine blades and tower are dynamically coupled and from an estimator design perspective, such interactions may provide an opportunity for the tower motion to be estimated based upon the blade load measurements that are already available to the IPC. If so, this indicates redundancy in the information provided by the tower motion sensor that can either be exploited in terms of a reduction in sensor count, or for fault tolerant control purposes (Feng, Patton, & Wang, 2014; Odgaard & Stoustrup, 2013; Sami & Patton, 2012). However, as will be shown in this chapter, estimating tower motion from blade loads is non-trivial, owing to the difference in respective frames of reference. Whilst the blade loads occur in a rotating frame of reference, the tower motions are with respect a stationary reference frame relative to the rotor. This gives rise to structural interactions that are of a periodically time-varying nature, complicating any subsequent observability analysis and observer design.

Such structural interactions are well known within the wind-energy community, where coordinate transformations such as the Coleman Transform (Bossanyi, 2003a; Markou, Buhl, Marrant, & van Engelen, 2006) are routinely employed to project quantities in rotational coordinates to a fixed frame of reference, and vice-versa.

As such, the Coleman Transform is of central importance to this work in terms of simplifying the wind-turbine model to one that is time-invariant. However, an important (and often overlooked) subtlety of the Coleman Transform is the fact that it is time-varying, and hence possesses dynamics. As was shown in Lu et al. (2015), the Coleman Transform introduces a frequency shifting effect upon rotating variables referred to a fixed frame of reference, and hence yields turbine models whose dynamics are significantly different to those that arise from a common misconception of the Coleman Transform being time-invariant.

The contributions are this chapter are thus twofold. Firstly, the observability of tower fore-aft motion is established using only standard blade-load measurements. This analysis is based upon a model that correctly accounts for the frequency shifting effects of the Coleman Transform and its inverse. Secondly, and with a view towards pragmatic adoption by industry, an observer-based tower damping controller is designed and implemented in a fashion that augments (rather than replaces) conventional pitch control architectures.

5.3 Modelling of Blade and Tower Dynamics

The wind turbine aerodynamic interactions of relevance to this study are depicted in Fig. 5.1. Owing to variable blade geometry, the wind-induced forces are not uniformly distributed on the blades and to model such forces blade element/momentum theory is adopted (Burton et al., 2011), where the blade is discretised into small elements. Referring to Fig. 5.1, the out-of-plane perturbation force $F_{i,l}^{\text{out}} : \mathbb{R} \times \mathbb{R} \times \mathbb{R} \to \mathbb{R}$ acting on each span-wise blade element $l \in \{1, \dots, L\} \subset \mathbb{Z}$ of blade $i \in \{1, 2, 3\}$ can be determined around the operating conditions according to the following relationship (van Engelen, 2006):

$$F_{i,l}^{\text{out}}(\theta_i, v_{i,l}, t) = \left.\frac{dF_{i,l}^{\text{out}}}{d\theta}\right|_{\theta^*, v^*} \theta_i(t) + \left.\frac{dF_{i,l}^{\text{out}}}{dv}\right|_{\theta^*, v^*} v_{i,l}(t), \tag{5.1}$$

where $\theta_i, v_{i,l} \in \mathbb{R}$ are the deviations of the blade pitch angle and apparent wind speed from their steady-states $\theta^*, v^* \in \mathbb{R}$, whilst $\frac{dF_{i,l}^{\text{out}}}{d\theta} \in \mathbb{R}$ and $\frac{dF_{i,l}^{\text{out}}}{dv} \in \mathbb{R}$ are the respective variations of out-of-plane force with respect to pitch angle and apparent wind speed.

Next, assuming the blade is rigid, the apparent stream-wise wind speed perturbation $v_{i,l} \in \mathbb{R}$ experienced by blade i on span-wise element l is dependent upon the actual stream-wise wind speed perturbation $v_{0_{i,l}} \in \mathbb{R}$, the fore-aft tower velocity $\dot{x}_{\text{fa}} \in \mathbb{R}$ and the rotational velocity $\dot{\varphi}_{\text{fa}} \in \mathbb{R}$, as follows:

$$v_{i,l}(t) = v_{0_{i,l}}(t) - \dot{x}_{\text{fa}}(t) + \dot{\varphi}_{\text{fa}}(t) r_l \sin\big(\phi_i(t)\big), \tag{5.2}$$

where $r_l \in \mathbb{R}$ is the radial distance of the l-th blade element. The azimuthal angle of each blade is defined as $[\phi_1(t), \phi_2(t), \phi_3(t)] := [\phi(t), \phi(t) + \frac{2\pi}{3}, \phi(t) + \frac{4\pi}{3}]$,

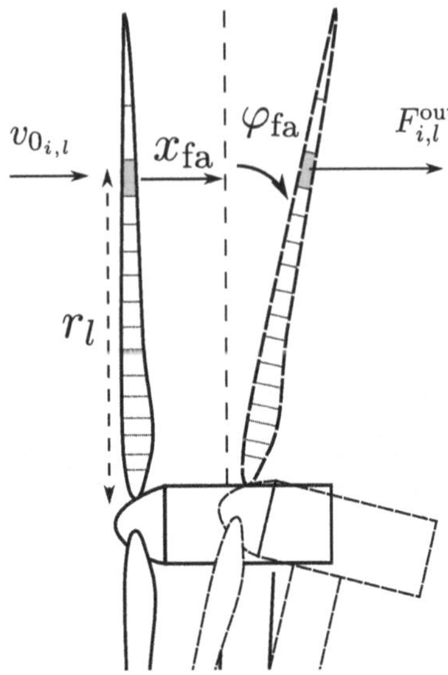

Fig. 5.1 An out-of-plane force $F_{i,l}^{out}$ on the shaded blade element at r_l is caused by the stream-wise wind speed $v_{0_{i,l}}$, fore-aft tower velocity $\dot{x}_{fa}$ and rotational velocity $\dot{\varphi}_{fa}$

where $\phi(t) \in \mathbb{R}$ is the angle of the first blade from the horizontal yaw axis with respect to the clockwise direction. This work implicitly assumes the tower is a prismatic beam so that the ratio between rotation and displacement is $\frac{2}{3h}$ (Selvam et al., 2009), where $h \in \mathbb{R}$ is the height of the tower. Thus, the fore-aft rotational velocity of the tower-top can be approximated as $\dot{\varphi}_{fa}(t) \approx \frac{2}{3h}\dot{x}_{fa}(t)$.

Assuming the wind-induced forces on the turbine hub are negligible in comparison to those on the blades and the aerofoil properties of the three turbine blades are identical, then:

$$\frac{dF_{i,l}^{out}}{d\theta} = \frac{dF_l^{out}}{d\theta}, \quad \frac{dF_{i,l}^{out}}{dv} = \frac{dF_l^{out}}{dv}, \quad \forall i = \{1, 2, 3\}. \tag{5.3}$$

Thus, the deviations of aerodynamic thrust F_a acting on the hub, and flap-wise blade root loading $M_{a_i} \in \mathbb{R}$ acting on the blades, from their steady-state are computed as follows:

$$F_a(t) = \sum_{i=1}^{3}\sum_{l=1}^{L} F_{i,l}^{out}(t), \tag{5.4a}$$

$$M_{a_i}(t) = \sum_{l=1}^{L} F_{i,l}^{out}(t) r_l. \tag{5.4b}$$

By substituting (5.1), (5.2) and (5.3) into (5.4a), the aerodynamic thrust $F_a(t)$ is simplified as follows:

$$
\begin{aligned}
F_{\mathrm{a}}(t) = & \Big(\theta_1(t) + \theta_2(t) + \theta_3(t)\Big) \sum_{l=1}^{L} \left[\frac{dF_l^{\mathrm{out}}}{d\theta} \right] \\
& + \sum_{i=1}^{3} \sum_{l=1}^{L} \left[\frac{dF_{i,l}^{\mathrm{out}}}{dv} \tilde{v}_{0_{i,l}}(t) \Big) \right] - \dot{x}_{\mathrm{fa}}(t) \sum_{i=1}^{3} \sum_{l=1}^{L} \left[\frac{dF_{i,l}^{\mathrm{out}}}{dv} \right] \\
& + \frac{2}{3h} \dot{x}_{\mathrm{fa}}(t) \sum_{i=1}^{3} \sum_{l=1}^{L} \left[\frac{dF_{i,l}^{\mathrm{out}}}{dv} r_l \sin\big(\phi_i(t)\big) \right].
\end{aligned} \tag{5.5}
$$

Thus, the aerodynamic thrust $F_{\mathrm{a}}(t) \in \mathbb{R}$ can be expressed as follows:

$$
F_{\mathrm{a}}(t) = \frac{dF_{\mathrm{a}}}{d\theta} \Big(\theta_1(t) + \theta_2(t) + \theta_3(t)\Big) + F_{\mathrm{a}}^{\mathrm{d}}(t) - k_{Fx} \dot{x}_{\mathrm{fa}}(t), \tag{5.6a}
$$

where $F_{\mathrm{a}}^{\mathrm{d}}(t) \in \mathbb{R}$ denotes the wind-induced tower disturbance and the remaining variables are defined as follows:

$$
\frac{dF_{\mathrm{a}}}{d\theta} := \sum_{l=1}^{L} \left[\frac{dF_l^{\mathrm{out}}}{d\theta} \right], \; F_{\mathrm{a}}^{\mathrm{d}}(t) := \sum_{i=1}^{3} \sum_{l=1}^{L} \left[\frac{dF_{i,l}^{\mathrm{out}}}{dv} v_{0_{i,l}}(t) \Big) \right], \tag{5.6b}
$$

$$
k_{Fx} := \sum_{i=1}^{3} \sum_{l=1}^{L} \left[\frac{dF_{i,l}^{\mathrm{out}}}{dv} \right], \; \sum_{i=1}^{3} \sum_{l=1}^{L} \left[\frac{dF_{i,l}^{\mathrm{out}}}{dv} r_l \sin\big(\phi_i(t)\big) \right] = 0. \tag{5.6c}
$$

Similarly, by substituting (5.1), (5.2) and (5.3) into (5.4b), the flap-wise blade root loading M_{a_i} is simplified as follows:

$$
\begin{aligned}
M_{\mathrm{a}_i}(t) = & \, \theta_i(t) \sum_{l=1}^{L} \left[\frac{dF_{i,l}^{\mathrm{out}}}{d\theta} r_l \right] + \sum_{l=1}^{L} \left[\frac{dF_{i,l}^{\mathrm{out}}}{dv} v_{0_{i,l}}(t) r_l \right] \\
& - \dot{x}_{\mathrm{fa}}(t) \sum_{l=1}^{L} \left[\frac{dF_{i,l}^{\mathrm{out}}}{dv} r_l \right] + \frac{2}{3h} \dot{x}_{\mathrm{fa}}(t) \sin\big(\phi_i(t)\big) \sum_{l=1}^{L} \left[\frac{dF_{i,l}^{\mathrm{out}}}{dv} r_l \right].
\end{aligned} \tag{5.7}
$$

Subsequently, the flap-wise blade root loading M_{a_i} can be expressed as follows:

$$
\begin{aligned}
M_{\mathrm{a}_i}(t) = & \frac{dM_{\mathrm{a}}}{d\theta} \theta_i(t) + M_i^{\mathrm{d}}(t) - k_{Mx} \dot{x}_{\mathrm{fa}}(t) \\
& + k_{M\varphi} \dot{x}_{\mathrm{fa}}(t) \sin\big(\phi_i(t)\big),
\end{aligned} \tag{5.8a}
$$

where $M_i^{\mathrm{d}}(t) \in \mathbb{R}$ is the wind-induced blade disturbance, with the terms above defined as:

$$\frac{dM_{\mathrm{a}}}{d\theta} := \sum_{l=1}^{L}\left[\frac{dF_{i,l}^{\mathrm{out}}}{d\theta} r_l\right],\ k_{Mx} := \sum_{l=1}^{L}\left[\frac{dF_{i,l}^{\mathrm{out}}}{dv} r_l\right], \tag{5.8b}$$

$$M_i^{\mathrm{d}}(t) := \sum_{l=1}^{L}\left[\frac{dF_{i,l}^{\mathrm{out}}}{dv} v_{0_{i,l}}(t) r_l\right],\ k_{M\varphi} := \frac{2}{3h} k_{Mx}. \tag{5.8c}$$

The wind-induced tower disturbance $F_{\mathrm{a}}^{\mathrm{d}}$ can be expressed in terms of the blade disturbances M_i^{d}. Considering (5.8c), the wind-induced blade disturbance M_i^{d} can be expressed as follows.

$$M_i^{\mathrm{d}}(t) = \sum_{l=1}^{L}\left[\frac{dF_{i,l}^{\mathrm{out}}}{dv} v_{0_{i,l}}(t) r_l\right] = \bar{F}_i^{\mathrm{d}}(t) \sum_{l=1}^{L} r_l, \tag{5.9a}$$

where $\bar{F}_i^{\mathrm{d}}(t) \in \mathbb{R}$ is the averaged wind-induced out-of-plane force on each blade element l along blade i, defined as:

$$\bar{F}_i^{\mathrm{d}}(t) := \frac{1}{L} \sum_{l=1}^{L}\left[\frac{dF_{i,l}^{\mathrm{out}}}{dv} v_{0_{i,l}}(t)\right]. \tag{5.9b}$$

Substituting (5.9b) into wind-induced tower disturbance $F_{\mathrm{a}}^{\mathrm{d}}$ from (5.6b) yields:

$$F_{\mathrm{a}}^{\mathrm{d}}(t) = \sum_{i=1}^{3}\sum_{l=1}^{L}\left[\frac{dF_{i,l}^{\mathrm{out}}}{dv} v_{0_{i,l}}(t)\right] = L \sum_{i=1}^{3} \bar{F}_i^{\mathrm{d}}(t). \tag{5.9c}$$

Collecting terms from (5.9a) and (5.9c) gives the desired result:

$$F_{\mathrm{a}}^{\mathrm{d}}(t) := k_{FM}\Big(M_1^{\mathrm{d}}(t) + M_2^{\mathrm{d}}(t) + M_3^{\mathrm{d}}(t)\Big), \tag{5.9d}$$

where $k_{FM} := L(\sum_{l=1}^{L} r_l)^{-1}$.

Finally, the dynamics of the blade root bending moments and fore-aft tower motion are approximated as second-order (e.g. van Engelen, 2006; Selvam et al., 2009) and are coupled as follows:

$$\ddot{M}_i(t) + 2\zeta_{\mathrm{b}}\omega_{\mathrm{b}}\dot{M}_i(t) + \omega_{\mathrm{b}}^2 M_i(t) = \omega_{\mathrm{b}}^2 M_{\mathrm{a}_i}(t), \tag{5.10a}$$

$$\ddot{x}_{\mathrm{fa}}(t) + 2\zeta_{\mathrm{t}}\omega_{\mathrm{t}}\dot{x}_{\mathrm{fa}}(t) + \omega_{\mathrm{t}}^2 x_{\mathrm{fa}}(t) = \frac{1}{m_{\mathrm{t}}}\big(F_{\mathrm{a}}(t) - \frac{2}{3h} M_{\mathrm{tilt}}(t)\big), \tag{5.10b}$$

where $M_i, x_{\mathrm{fa}} \in \mathbb{R}$ denote the deviations of the flap-wise ith blade root bending moment and tower fore-aft displacement from equilibrium, respectively. The damping ratio of the blade and tower are $\zeta_{\mathrm{b}}, \zeta_{\mathrm{t}} \in \mathbb{R}$, respectively, m_{t} is the mass of the tower-top and $\omega_{\mathrm{b}}, \omega_{\mathrm{t}} \in \mathbb{R}$ are the respective natural frequencies of the blade and tower. The tilt moment of the rotor at the tower-top is defined as $M_{\text{tilt}}(t) = \frac{2}{3}\sum_{i=1}^{3} M_i(t)\sin(\phi_i(t))$ (Selvam et al., 2009).

Inspection of (5.6), (5.8) and (5.10) indicates the coupling between the blade bending moments and the tower dynamics, which is the key property that underpins the subsequent work in this chapter.

The state-space representation of (5.10) can be formulated as follows:

$$\begin{aligned} \dot{x}(t) &= A(t)x(t) + Bu(t) + B_{\text{d}}d(t), \\ y(t) &= Cx(t), \end{aligned} \tag{5.11a}$$

with

$$x(t) := \left[\dot{\mathbf{M}}(t)\ \mathbf{M}(t)\ \dot{x}_{\text{fa}}(t)\ x_{\text{fa}}(t)\right]^T \in \mathbb{R}^{n_x}, \tag{5.11b}$$

$$u(t) := \left[\theta_1(t)\ \theta_2(t)\ \theta_3(t)\right]^T \in \mathbb{R}^{n_u}, \tag{5.11c}$$

$$d(t) := \left[M_1^{\text{d}}(t)\ M_2^{\text{d}}(t)\ M_3^{\text{d}}(t)\right]^T \in \mathbb{R}^{n_d}, \tag{5.11d}$$

$$y(t) := \mathbf{M}(t) \in \mathbb{R}^{n_y}, \tag{5.11e}$$

$$A(t) := \begin{bmatrix} -2\zeta_{\text{b}}\omega_{\text{b}} I & -\omega_{\text{b}}^2 I & -k_{Mx}\mathbf{1} + k_{M\varphi}\mathbf{S}(\phi(t)) & 0 \\ I & 0 & 0 & 0 \\ 0 & \frac{-4}{9hm_{\text{t}}}\mathbf{S}^T(\phi(t)) & -2\zeta_{\text{t}}\omega_{\text{t}} - \frac{k_{Fx}}{m_{\text{t}}} & -\omega_{\text{t}}^2 \\ 0 & 0 & 1 & 0 \end{bmatrix}, \tag{5.11f}$$

$$B := \begin{bmatrix} \omega_{\text{b}}^2 \frac{dM_{\text{a}}}{d\theta} I \\ 0 \\ \frac{1}{m_{\text{t}}}\frac{dF_{\text{a}}}{d\theta}\mathbf{1}^T \\ 0 \end{bmatrix}, \quad B_{\text{d}} := \begin{bmatrix} \omega_{\text{b}}^2 I \\ 0 \\ \frac{1}{m_{\text{t}}} k_{FM}\mathbf{1}^T \\ 0 \end{bmatrix}, \quad C := \begin{bmatrix} 0 \\ I \\ 0 \\ 0 \end{bmatrix}, \tag{5.11g}$$

$$\mathbf{M}(t) := \begin{bmatrix} M_1(t) \\ M_2(t) \\ M_3(t) \end{bmatrix}, \mathbf{S}(\phi(t)) := \begin{bmatrix} \sin(\phi(t)) \\ \sin\left(\phi(t) + \frac{2\pi}{3}\right) \\ \sin\left(\phi(t) + \frac{4\pi}{3}\right) \end{bmatrix}, \mathbf{1} := \begin{bmatrix} 1 \\ 1 \\ 1 \end{bmatrix}. \tag{5.11h}$$

The numeric values of the model parameters employed in this work are provided in Table 5.1. Notice that $\phi(t) = \omega(t)t$, where $\omega(t)$ is the rotor speed. Under the

Table 5.1 Model parameters

Parameters	Values	Units	Parameters	Values	Units
ω_{t}	2.01	rad s^{-1}	ω_{b}	4.40	rad s^{-1}
ζ_{t}	1.00	–	ζ_{b}	0.47	–
$\frac{dF_{\text{a}}}{d\theta}$	−48.29	kN deg^{-1}	$\frac{dM_{\text{a}}}{d\theta}$	−1.50	MNm deg^{-1}
k_{Fx}	55.72	kNsm^{-1}	k_{Mx}	1.35	MNs
m_{t}	34.75×10^4	kg			

additional assumption that the rotor operates at a rated speed $\omega(t) = \omega_0$ in above-rated conditions, then $\phi(t) = \omega_0 t$ and the system (5.11) becomes a linear time-periodic system with a period of $T = \frac{2\pi}{\omega_0}$.

5.4 Analysis of the Time-Periodic System Observability

For estimator design, it is necessary to assess the observability of the system. This section examines the observability properties of the linear time-periodic system (5.11).

5.4.1 Preliminaries on Observability and Linear Periodic Systems

This section recalls a few definitions and theorems pertinent to linear time-periodic systems.

Definition 5.4.1 (*Linear time-periodic system*) A linear time-periodic system is described as follows:

$$\dot{x}(t) = A(t)x(t) + Bu(t), \; y(t) = Cx(t), \tag{5.12}$$

with state $x \in \mathbb{R}^{n_x}$, input $u \in \mathbb{R}^{n_u}$, output $y \in \mathbb{R}^{n_y}$ and $A(t) \in \mathbb{R}^{n_x \times n_x}$ is periodic with period T, namely $A(t) = A(t + T)$.

According to Bittanti and Colaneri (2009), for the system (5.12), there exists a state transition matrix $\Phi(t, t_0) \in \mathbb{R}^{n_x \times n_x}$ such that:

$$x(t) = \Phi(t, t_0)x(t_0). \tag{5.13}$$

Theorem 5.1 (Brockett, 1970). *The linear time-periodic system* (5.12) *is asymptotically stable if and only if the eigenvalues of the state transition matrix* $\Phi(T, 0)$ *lie within the unit circle.*

Definition 5.4.2 (Bittanti & Colaneri, 2009) The observability Gramian of (5.12) is:

$$W_o(t_0, t_f) = \int_{t_0}^{t_f} \Phi^T(t, t_0)C^T C\Phi(t, t_0)dt. \tag{5.14}$$

Theorem 5.2 (D'Angelo, 1970) *The system* (5.12) *is observable over the time interval* $[t_0, t_f]$ *if and only if* $W_o(t_0, t_f)$ *is positive definite.*

5.4.2 Observability of the Time-Periodic Systems

Lemma 5.3 *Assume a constant rotor speed* $\omega(t) = \omega_0$, *the linear periodic system (5.11) is observable over the interval* $[t_0, t_f]$.

Proof To examine the observability of (5.11), from Theorem 5.2, the observability Gramian $W_o(t_0, t_f)$ of (5.11):

$$W_0(t_0, t_f) = \int_{t_0}^{t_f} \Phi^T(t, t_0) C^T C^T \Phi(t, t_0) dt, \tag{5.15}$$

needs to be positive definite. However, finding the analytical expression of $\Phi(t, t_0)$ and $W_0(t_0, t_f)$ is not trivial for time-varying systems like (5.11). Nonetheless, there is a theorem proposed by Chen (1984) that can examine the observability without computing the state transition. Assume $A(t) \in \mathbb{R}^{n_x \times n_x}$ and $C \in \mathbb{R}^{n_y \times n_x}$ are $q-1$ and q times continuously differentiable, respectively, and consider a matrix defined as follows:

$$N(t) = [N_0(t), \ldots, N_q(t)]^T, \tag{5.16a}$$

where

$$N_0(t) = C, \tag{5.16b}$$

$$N_{m+1}(t) = N_m(t) A(t) + \dot{N}_m(t), \; m = 0, 1, 2, \ldots, q. \tag{5.16c}$$

If $N(t)$, where $t \in [t_0, t_f]$, has rank n_x, then $W_0(t_0, t_f)$ is positive definite (Chen, 1984). Consider $q = 3$, $N(t)$ becomes:

$$N(t) = \begin{bmatrix} 0 & I & 0 & 0 \\ I & 0 & 0 & 0 \\ -2\omega_b\zeta_b I & -\omega_b^2 I & N^{(3,3)}(t) & 0 \\ (4\omega_b^2\zeta_b^2 - \omega_b^2)I & N^{(4,2)}(t) & N^{(4,3)}(t) & N^{(4,4)}(t) \end{bmatrix}, \tag{5.17a}$$

where

$$N^{(3,3)}(t) = k_{M\varphi}\mathbf{S}(\omega_0 t) - k_{Mx}\mathbf{1}, \tag{5.17b}$$

$$N^{(4,2)}(t) = 2\zeta_b\omega_b^3 I - \frac{3}{2h}\mathbf{S}(\phi(t))\Big(k_{Mx}\mathbf{1} - k_{M\varphi}\mathbf{S}^T(\phi(t))\Big), \tag{5.17c}$$

$$N^{(4,3)}(t) = \big(k_{Fx}\omega_t^2 + 2\zeta_t\omega_t + 2\zeta_b\omega_b\big) \times \Big(k_{Mx}\mathbf{1} - k_{M\varphi}\mathbf{S}(\phi(t))\Big) + k_{M\varphi}\dot{\mathbf{S}}(\phi(t)), \tag{5.17d}$$

$$N^{(4,4)}(t) = \omega_t^2\Big(k_{Mx}\mathbf{1} - k_{M\varphi}\mathbf{S}(\phi(t))\Big). \tag{5.17e}$$

Since $N^{(3,3)}(t)$ and $N^{(4,4)}(t)$ both are full rank for all t, thus, $N(t)$ has rank n_x over $t \in [t_0, t_f]$. Therefore, the system (5.11) is observable. □

Lemma 5.3 indicates that use of observers can reconstruct the tower disturbance based on the linear time-periodic model (5.11). However, it is non-trivial to design an observer based on a linear time-periodic models, thus, the model (5.11) can be transformed into a time-invariant model, as discussed in the following section.

5.5 Transformation to an LTI System and Observability Analysis

Under the assumption of a constant rotor speed, the system model (5.11) is a linear, time-periodic (LTP) system (D'Angelo, 1970) and for such LTP systems, there exist techniques for observability analysis and estimator design (e.g. Montagnier, Spiteri, & Angeles, 2004). However, the problem of establishing the observability proof and synthesising an estimator for the LTP system (5.11) can be greatly simplified by reformulating (5.11) as a LTI system. As will now be shown, the key to achieving this lies in the use of a coordinate transformation based upon the Coleman Transform.

The Coleman Transform projects the blade loads in the rotating frame of reference onto the fixed tilt and yaw turbine axes, and play a key role in transforming (5.11) into an LTI system. The typical Coleman transform $T_{\text{cm}}\left(\phi(t)\right) \in \mathbb{R}^{3\times 3}$ is defined as follows (e.g. Lu et al., 2015 and references therein):

$$\begin{bmatrix} \bar{M}(t) \\ M_{\text{tilt}}(t) \\ M_{\text{yaw}}(t) \end{bmatrix} = \underbrace{\frac{2}{3}\begin{bmatrix} \frac{1}{2} & \frac{1}{2} & \frac{1}{2} \\ \sin(\phi(t)) & \sin\left(\phi(t)+\frac{2\pi}{3}\right) & \sin\left(\phi(t)+\frac{4\pi}{3}\right) \\ \cos(\phi(t)) & \cos\left(\phi(t)+\frac{2\pi}{3}\right) & \cos\left(\phi(t)+\frac{4\pi}{3}\right) \end{bmatrix}}_{T_{\text{cm}}\left(\phi(t)\right)} \times \begin{bmatrix} M_1(t) \\ M_2(t) \\ M_3(t) \end{bmatrix}, \tag{5.18a}$$

where $\bar{M}$, M_{tilt}, $M_{\text{yaw}} \in \mathbb{R}$ denote the collective, tilt and yaw referred flap-wise blade root bending moments, respectively. The inverse Coleman transform $T_{\text{cm}}^{\text{inv}}\left(\phi(t)\right) \in \mathbb{R}^{3\times 3}$ is as follows:

$$\begin{bmatrix} M_{\text{a}_1}(t) \\ M_{\text{a}_2}(t) \\ M_{\text{a}_3}(t) \end{bmatrix} = \underbrace{\begin{bmatrix} 1 & \sin(\phi(t)) & \cos(\phi(t)) \\ 1 & \sin\left(\phi(t)+\frac{2\pi}{3}\right) & \cos\left(\phi(t)+\frac{2\pi}{3}\right) \\ 1 & \sin\left(\phi(t)+\frac{4\pi}{3}\right) & \cos\left(\phi(t)+\frac{4\pi}{3}\right) \end{bmatrix}}_{T_{\text{cm}}^{\text{inv}}\left(\phi(t)\right)} \times \begin{bmatrix} \bar{M}_{\text{a}}(t) \\ M_{\text{a}_{\text{tilt}}}(t) \\ M_{\text{a}_{\text{yaw}}}(t) \end{bmatrix}, \tag{5.18b}$$

where $\bar{M}_{\text{a}}$, $M_{\text{a}_{\text{tilt}}}$, $M_{\text{a}_{\text{yaw}}} \in \mathbb{R}$ represent the collective, tilt and yaw referred aerodynamic forces upon a non-rotating reference frame, respectively. Clearly, the Coleman Transforms are time dependent, and hence their dynamics must be factored into any system model that employs them. As shown in Lu et al. (2015) from the perspective of IPC design, models that arise from misconceived treatment of the Coleman Transforms as static projections give rise to erroneous dynamics, leading to poor IPC performance. Now, the following theorem shows the main result of this chapter, which is the LTI reformulation of (5.11) with the correct treatment of the Coleman Transforms.

Theorem 5.4 (Main result) *Assuming a fixed rotor speed and Coleman transformations (5.18), the linear time-varying system (5.11) can be transformed into the following linear time-invariant form:*

$$\begin{bmatrix} \dot{\xi}(t) \\ \dot{x}_{\rm t}(t) \end{bmatrix} = \underbrace{\begin{bmatrix} A_\xi & B_{\xi \rm t} \\ B_{\rm t\xi} & A_{\rm t} \end{bmatrix}}_{A_{\rm lti}} \begin{bmatrix} \xi(t) \\ x_{\rm t}(t) \end{bmatrix} + \begin{bmatrix} B_{\xi\theta} \\ B_{\rm t\theta} \end{bmatrix} u_{\rm cm}(t) + \begin{bmatrix} B_{\xi \rm d} \\ B_{\rm td} \end{bmatrix} d_{\rm cm}(t),$$

$$y_{\rm cm}(t) = \underbrace{\begin{bmatrix} C_\xi & 0 \end{bmatrix}}_{C_{\rm lti}} \begin{bmatrix} \xi(t) \\ x_{\rm t}(t) \end{bmatrix}, \tag{5.19a}$$

where $\xi \in \mathbb{R}^{n_\xi}$ is the projection of the states associated with the blade dynamics upon a non-rotating reference frame (5.27), $x_{\rm t} \in \mathbb{R}^{n_{x_{\rm t}}}$ are the states of the tower dynamics (5.28), and the referred measurements of flap-wise blade moments, pitch angle signals and wind-induced blade disturbance upon a non-rotating coordinate frame are as follows:

$$y_{\rm cm}(t) := \left[\bar{M}(t), M_{\rm tilt}(t), M_{\rm yaw}(t)\right]^T \in \mathbb{R}^{n_y}, \tag{5.19b}$$

$$u_{\rm cm}(t) := \left[\bar{\theta}(t), \theta_{\rm tilt}(t), \theta_{\rm yaw}(t)\right]^T \in \mathbb{R}^{n_u}, \tag{5.19c}$$

$$d_{\rm cm}(t) := \left[\bar{M}^{\rm d}(t), M^{\rm d}_{\rm tilt}(t), M^{\rm d}_{\rm yaw}(t)\right]^T \in \mathbb{R}^{n_d}. \tag{5.19d}$$

Proof The proof uses the following properties:

$$\begin{aligned} \mathcal{L}[u(t)\sin\phi(t)] &= \mathcal{L}\left[u(t)\frac{j(e^{-j\omega_0 t} - e^{j\omega_0 t})}{2}\right], \\ &= \frac{j}{2}\big(u(s + j\omega_0) - u(s - j\omega_0)\big), \end{aligned} \tag{5.20a}$$

$$\begin{aligned} \mathcal{L}\left[u(t)\cos\phi(t)\right] &= \mathcal{L}\left[u(t)\frac{e^{j\omega_0 t} + e^{-j\omega_0 t}}{2}\right], \\ &= \frac{1}{2}\big(u(s - j\omega_0) + u(s + j\omega_0)\big), \end{aligned} \tag{5.20b}$$

where $u(t)$ is an arbitrary input signal and $u(s)$ is its Laplace transform. Substituting identities (5.20) into Coleman transformations (5.18) yields:

$$\begin{bmatrix} \bar{M}(s) \\ M_{\text{tilt}}(s) \\ M_{\text{yaw}}(s) \end{bmatrix} = \frac{2}{3}C_- \begin{bmatrix} M_1(s - j\omega_0) \\ M_2(s - j\omega_0) \\ M_3(s - j\omega_0) \end{bmatrix} + \frac{2}{3}C_+ \begin{bmatrix} M_1(s + j\omega_0) \\ M_2(s + j\omega_0) \\ M_3(s + j\omega_0) \end{bmatrix} + \frac{1}{3}C_0 \begin{bmatrix} M_1(s) \\ M_2(s) \\ M_3(s) \end{bmatrix}, \tag{5.21a}$$

$$\begin{bmatrix} M_{a_1}(s) \\ M_{a_2}(s) \\ M_{a_3}(s) \end{bmatrix} = C_-^T \begin{bmatrix} \bar{M}_a(s - j\omega_0) \\ M_{a_{\text{tilt}}}(s - j\omega_0) \\ M_{a_{\text{yaw}}}(s - j\omega_0) \end{bmatrix} + C_+^T \begin{bmatrix} \bar{M}_a(s + j\omega_0) \\ M_{a_{\text{tilt}}}(s + j\omega_0) \\ M_{a_{\text{yaw}}}(s + j\omega_0) \end{bmatrix} + C_0^T \begin{bmatrix} \bar{M}_a(s) \\ M_{a_{\text{tilt}}}(s) \\ M_{a_{\text{yaw}}}(s) \end{bmatrix}, \tag{5.21b}$$

where C_-, C_+ and C_0 are defined as:

$$C_- := \frac{1}{2}\begin{bmatrix} 0 & 0 & 0 \\ 0 & 1 & -j \\ 0 & j & 1 \end{bmatrix} \begin{bmatrix} 0 & 0 & 0 \\ \sin(0) & \sin(\frac{2\pi}{3}) & \sin(\frac{4\pi}{3}) \\ \cos(0) & \cos(\frac{2\pi}{3}) & \cos(\frac{4\pi}{3}) \end{bmatrix}, \tag{5.21c}$$

$$C_+ := \frac{1}{2}\begin{bmatrix} 0 & 0 & 0 \\ 0 & 1 & j \\ 0 & -j & 1 \end{bmatrix} \begin{bmatrix} 0 & 0 & 0 \\ \sin(0) & \sin(\frac{2\pi}{3}) & \sin(\frac{4\pi}{3}) \\ \cos(0) & \cos(\frac{2\pi}{3}) & \cos(\frac{4\pi}{3}) \end{bmatrix}, \tag{5.21d}$$

$$C_0 := \begin{bmatrix} 1 & 1 & 1 \\ 0 & 0 & 0 \\ 0 & 0 & 0 \end{bmatrix}. \tag{5.21e}$$

Consider the blade model upon a rotating frame of reference (5.10a) and its Laplace transform:

$$M_i(s) = G(s)M_{a_i}(s), \tag{5.22}$$

where $G(s) = C_b(sI - A_b)^{-1}B_b$, with $A_b \in \mathbb{R}^{n_b \times n_b}$, $B_b \in \mathbb{R}^{n_b}$, $C_b \in \mathbb{R}^{1 \times n_b}$. Substituting (5.22) into (5.21) yields the following Coleman-transformed model in a fixed reference frame:

$$\begin{bmatrix} \bar{M}(s) \\ M_{\text{tilt}}(s) \\ M_{\text{yaw}}(s) \end{bmatrix} = \begin{bmatrix} G(s) & 0 & 0 \\ 0 & G_+(s) & G_-(s) \\ 0 & -G_-(s) & G_+(s) \end{bmatrix} \begin{bmatrix} \bar{M}_a(s) \\ M_{a_{\text{tilt}}}(s) \\ M_{a_{\text{yaw}}}(s) \end{bmatrix}, \tag{5.23a}$$

where $G_+, G_- \in \mathcal{R}$ are defined as:

$$G_+(s) := \frac{G(s + j\omega_0) + G(s - j\omega_0)}{2}, \tag{5.23b}$$

$$G_-(s) := j\frac{G(s + j\omega_0) - G(s - j\omega_0)}{2}, \tag{5.23c}$$

and $G(s + j\omega_0) = C_\text{b}(sI - (A_\text{b} - j\omega_0 I))^{-1}B_\text{b}$ and $G(s - j\omega_0) = C_\text{b}(sI - (A_\text{b} + j\omega_0 I))^{-1}B_\text{b}$. Subsequently, the Coleman transformed model (5.23) can be expressed in a state-space form, with state $x_\text{cm} \in \mathbb{C}^{5n_\text{b}}$, as follows:

$$\dot{x}_\text{cm}(t) = A_\text{cm}x_\text{cm}(t) + B_\text{cm}M_{\text{a}_\text{cm}}(t), \ y_\text{cm}(t) = C_\text{cm}x_\text{cm}(t), \tag{5.24a}$$

with $A_\text{cm} \in \mathbb{C}^{5n_\text{b}\times 5n_\text{b}}$, $B_\text{cm} \in \mathbb{C}^{5n_\text{b}\times 3}$, $C_\text{cm} \in \mathbb{R}^{3\times 5n_\text{b}}$ given as:

$$A_\text{cm} := \begin{bmatrix} A_\text{b} & 0 & 0 & 0 & 0 \\ 0 & A_\text{b} - j\omega_0 I & 0 & 0 & 0 \\ 0 & 0 & A_\text{b} + j\omega_0 I & 0 & 0 \\ 0 & 0 & 0 & A_\text{b} - j\omega_0 I & 0 \\ 0 & 0 & 0 & 0 & A_\text{b} + j\omega_0 I \end{bmatrix}, \tag{5.24b}$$

$$B_\text{cm} := \begin{bmatrix} B_\text{b} & 0 & 0 \\ 0 & B_\text{b} & jB_\text{b} \\ 0 & B_\text{b} & -jB_\text{b} \\ 0 & -jB_\text{b} & B_\text{b} \\ 0 & jB_\text{b} & B_\text{b} \end{bmatrix}, \ C_\text{cm}^T := \begin{bmatrix} C_\text{b} & 0 & 0 \\ 0 & \frac{1}{2}C_\text{b} & 0 \\ 0 & \frac{1}{2}C_\text{b} & 0 \\ 0 & 0 & \frac{1}{2}C_\text{b} \\ 0 & 0 & \frac{1}{2}C_\text{b} \end{bmatrix}, \tag{5.24c}$$

and where M_{a_cm} is obtained by substituting the Coleman transform (5.18) into (5.8):

$$M_{a_\text{cm}}(t) = \frac{dM_\text{a}}{d\theta}u_\text{cm}(t) + d_\text{cm}(t) + k_{Mx_\text{cm}}\dot{x}_\text{fa}(t), \tag{5.24d}$$

with $k_{Mx_\text{cm}} := \left[-k_{Mx}, k_{M\varphi}, 0\right]^T$. The LTI model (5.24) is not real-valued but the use of a state transformation matrix $T \in \mathbb{C}^{5n_\text{b}\times 5n_\text{b}}$ can map $x_\text{cm} \in \mathbb{C}^{5n_\text{b}}$ into $\xi = Tx_\text{cm} \in \mathbb{R}^{5n_\text{b}}$, where:

$$T := \begin{bmatrix} 1 & 0 & 0 \\ 0 & T_\text{c} & 0 \\ 0 & 0 & T_\text{c} \end{bmatrix}, \ T_\text{c} := \frac{1}{2}\begin{bmatrix} (1+j) & (1-j) \\ (1-j) & (1+j) \end{bmatrix}. \tag{5.25}$$

The equivalent model of (5.24) with the real-valued state ξ is:

$$\dot{\xi}(t) = A_\xi\xi(t) + B_\xi M_{\text{a}_\text{cm}}(t), \ y_\text{cm}(t) = C_\xi\xi(t), \tag{5.26a}$$

where

$$A_\xi = TA_{\text{cm}}T^{-1} = \begin{bmatrix} A_\text{b} & 0 & 0 & 0 & 0 \\ 0 & A_\text{b} & -\omega_0 I & 0 & 0 \\ 0 & \omega_0 I & A_\text{b} & 0 & 0 \\ 0 & 0 & 0 & A_\text{b} & -\omega_0 I \\ 0 & 0 & 0 & \omega_0 I & A_\text{b} \end{bmatrix}, \tag{5.26b}$$

$$B_\xi = TB_{\text{cm}} = \begin{bmatrix} B_\text{b} & 0 & 0 \\ 0 & B_\text{b} & B_\text{b} \\ 0 & B_\text{b} & -B_\text{b} \\ 0 & -B_\text{b} & B_\text{b} \\ 0 & B_\text{b} & B_\text{b} \end{bmatrix}, \; C_\xi = C_{\text{cm}}T^{-1} = C_{\text{cm}}. \tag{5.26c}$$

By substituting $M_{a_{\text{cm}}}$ from (5.24d) into (5.26), the model (5.26) becomes:

$$\dot{\xi}(t) = A_\xi \xi(t) + B_{\xi\theta} u_{\text{cm}}(t) + B_{\xi d} d_{\text{cm}}(t) + B_{\xi t} x_\text{t}(t), \tag{5.27a}$$

$$y_{\text{cm}}(t) = C_\xi \xi(t), \tag{5.27b}$$

where

$$B_{\xi\theta} := B_\xi \frac{dM_\text{a}}{d\theta}, \; B_{\xi d} := B_\xi, \; B_{\xi t} := B_\xi k_{Mx_{\text{cm}}} \begin{bmatrix} 1 & 0 \end{bmatrix}, \tag{5.27c}$$

and $x_\text{t} \in \mathbb{R}^{n_{x_\text{t}}}$ is the state of the state-space form of the tower dynamics (5.10b), that is defined as follows:

$$\begin{bmatrix} \ddot{x}_{\text{fa}}(t) \\ \dot{x}_{\text{fa}}(t) \end{bmatrix} = \underbrace{\begin{bmatrix} -2\omega_\text{t}\zeta_\text{t} & -\omega_\text{t}^2 \\ 1 & 0 \end{bmatrix}}_{A_\text{t}} \underbrace{\begin{bmatrix} \dot{x}_{\text{fa}}(t) \\ x_{\text{fa}}(t) \end{bmatrix}}_{x_\text{t}(t)} + \underbrace{\begin{bmatrix} \frac{dF_\text{a}}{d\theta} E_1 \\ 0 \end{bmatrix}}_{B_{\text{t}\theta}} u_{\text{cm}}(t)$$
$$+ \underbrace{\begin{bmatrix} k_{FM} E_1 \\ 0 \end{bmatrix}}_{B_{\text{td}}} d_{\text{cm}}(t) + \underbrace{\begin{bmatrix} \frac{-2}{3hm_\text{t}} C_\xi^{(2,:)} \\ 0 \end{bmatrix}}_{B_{\text{t}\xi}} \xi(t). \tag{5.28}$$

where $E_1 = [\frac{1}{m_\text{t}}, 0, 0]$ and $C_\xi^{(2,:)}$ denotes the second row of C_ξ. Finally, augmenting (5.26) with the tower dynamics (5.28) yields (5.19). □

Corollary 5.5.1 *Assuming the model parameters listed in Table 5.1, the system (5.19) is observable.*

Proof The observability proof of a continuous-time LTI system can be established if its observability matrix has full rank (Kalman, 1959). The observability matrix $\mathcal{O}$ of (5.19) is defined as follows:

$$\mathcal{O} = \begin{bmatrix} C_{\text{lit}} & C_{\text{lti}} A_{\text{lti}} & \cdots & C_{\text{lti}} A_{\text{lti}}^{n_\xi + n_{x_\text{t}} - 1} \end{bmatrix}^T. \tag{5.29}$$

If the system (5.19) is constructed based on the parameters listed in Table 5.1, then the observability matrix $\mathcal{O}$ of (5.19) has full rank $n_\xi + n_{x_t}$. Thus, the system (5.19) is observable. □

Hence, the tower motion states are observable from measurements of the blade loads alone. This result lays the foundation for the observer and controller designs of the next section.

5.6 Design of the Tower Estimator and Controller

Figure 5.2 depicts the architecture of the proposed control system, where the tower motion estimator produces an estimate $\hat{\dot{x}}_{\text{fa}}$ of the fore-aft velocity of the tower-top, based on the (Coleman-transformed) blade moment measurements $\bar{M}$, M_{tilt}, M_{yaw} and pitch signals $\bar{\theta}$, θ_{tilt}, θ_{yaw}. The tower controller subsequently employs this estimate to provides additional collective blade pitch signals $\bar{\theta}$ for attenuating the tower motion. Note that this architecture is deliberately chosen so as to augment, rather than replace the existing CPC and IPC controllers.

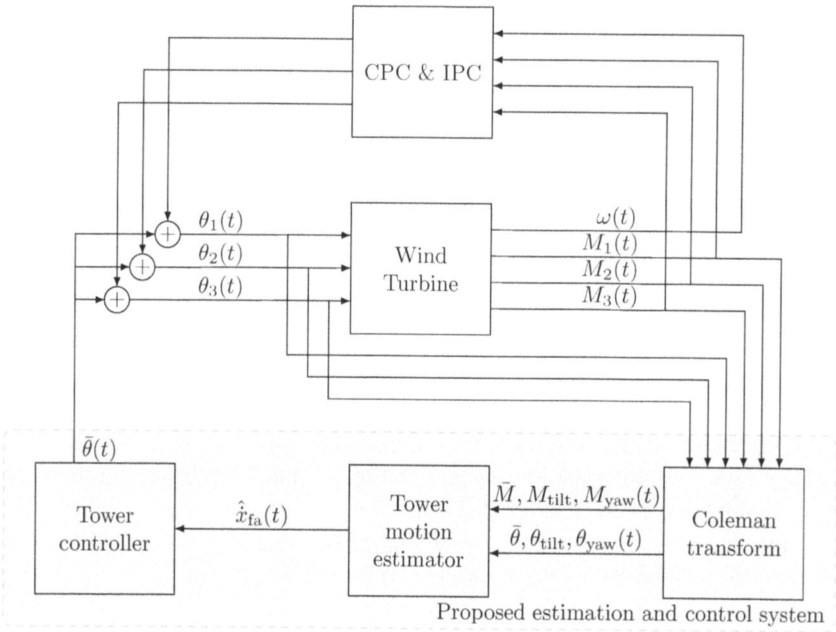

Fig. 5.2 Schematic of the proposed estimator and controller

5.6.1 Estimator Design

The estimator employed in this work is an unknown input disturbance observer (Johnson, 1971) that uses the modelled system (5.19) augmented with a wind-induced disturbance model. For brevity, a constant wind-induced disturbance model is assumed (e.g. Selvam et al., 2009):

$$\dot{d}_{\mathrm{cm}}(t) = 0, \tag{5.30}$$

Thus, the augmented model is described as follows:

$$\begin{bmatrix} \dot{\hat{\xi}}(t) \\ \dot{\hat{x}}_{\mathrm{t}}(t) \\ \dot{\hat{d}}_{\mathrm{cm}}(t) \end{bmatrix} = \underbrace{\begin{bmatrix} A_\xi & B_{\xi\mathrm{t}} & B_{\xi\mathrm{d}} \\ B_{\mathrm{tM}} & A_{\mathrm{t}} & B_{\mathrm{td}} \\ 0 & 0 & 0 \end{bmatrix}}_{A_{\mathrm{aug}}} \underbrace{\begin{bmatrix} \hat{\xi}(t) \\ \hat{x}_{\mathrm{t}}(t) \\ \hat{d}_{\mathrm{cm}}(t) \end{bmatrix}}_{\hat{z}(t)} + \underbrace{\begin{bmatrix} B_{\xi\theta} \\ B_{\mathrm{t}\theta} \\ 0 \end{bmatrix}}_{B_{\mathrm{aug}}} u_{\mathrm{cm}}(t)$$
$$+ L_{\mathrm{o}}\Big(y_{\mathrm{cm}}(t) - \underbrace{\begin{bmatrix} C_\xi & 0 & 0 \end{bmatrix}}_{C_{\mathrm{aug}}} \hat{z}(t)\Big), \tag{5.31}$$

where $\hat{z} \in \mathbb{R}^{n_z}$ and the hat symbol denotes estimate, whilst L_{o} represents the estimator gain. This system is also observable, and the estimator gain can be optimised by Kalman filtering theory (Kalman, 1960).

5.6.2 Estimation-Based Controller Design

Typically, a tower controller adjusts the blade pitch collectively in response to the tower fore-aft velocity in order to attenuate the harmful tower loads, where such loads are caused by symmetric loadings on the rotor and mainly concentrated around the resonant frequency of the tower (0.32 Hz in this work) (Burton et al., 2011). In contrast, a CPC also utilises the collective pitch angle to regulate the rotor speed at frequencies below the rotational frequency of the rotor (e.g. 0.2 Hz), whilst an IPC targets the blade loads at harmonics of the blade rotational frequency (e.g. 0.2, 0.4 Hz). Thus, assuming the architecture of Fig. 5.3, the tower controller must be designed not to excite the CPC or IPC loops.

Referring to the more detailed closed-loop system diagram of Fig. 5.3, the relationship between turbine inputs and outputs is as follows:

$$\begin{bmatrix} \omega(s) \\ \bar{M}(s) \\ M_{\mathrm{tilt}}(s) \\ M_{\mathrm{yaw}}(s) \end{bmatrix} = \begin{bmatrix} \begin{bmatrix} G_{\mathrm{cpc}}(s) & 0 & 0 \end{bmatrix} \\ G_{\mathrm{cm}}(s) \end{bmatrix} \begin{bmatrix} \bar{\theta}(s) \\ \theta_{\mathrm{tilt}}(s) \\ \theta_{\mathrm{yaw}}(s) \end{bmatrix}, \tag{5.32}$$

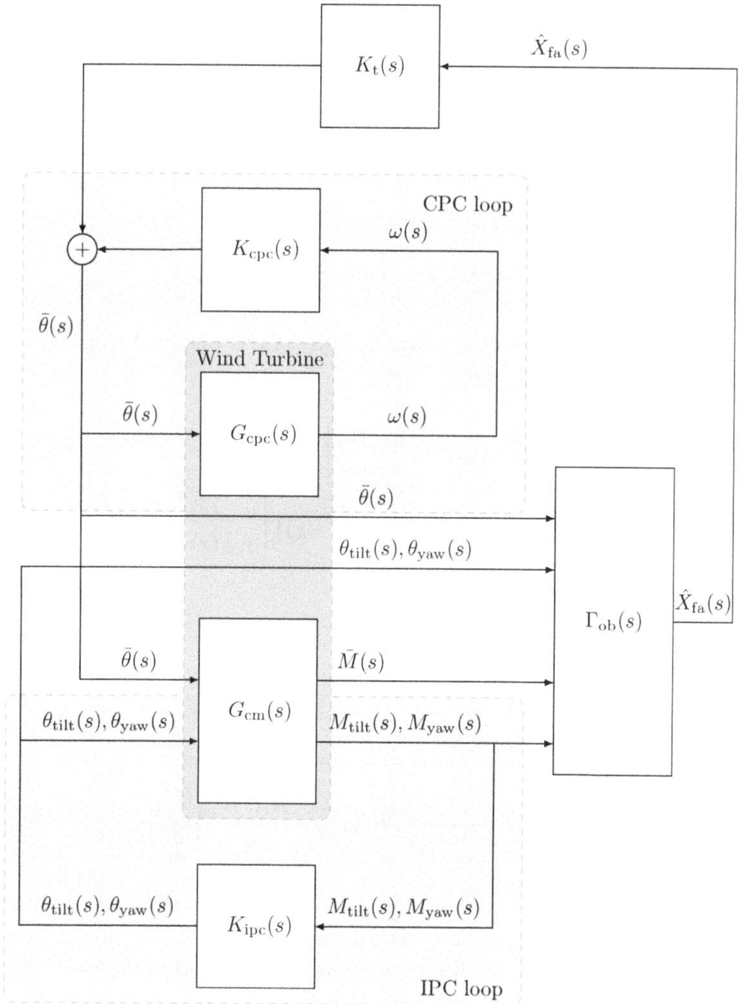

Fig. 5.3 System architecture for the tower controller design

where the transfer function mapping $u_{cm} = [\bar{\theta}, \theta_{tilt}, \theta_{yaw}]^T$ to $y_{cm} = [\bar{M}, M_{tilt}, M_{yaw}]^T$ is denoted $G_{cm} \in \mathcal{R}^{3\times3}$, and is obtained from (5.19). The transfer function of the rotor dynamics relating collective pitch inputs to rotor speed output is denoted $G_{cpc} \in \mathcal{R}$, that is discussed in Sect. 2.8 in Chap. 2. In terms of the controller inputs and outputs, the relationships are as follows:

$$\begin{bmatrix} \bar{\theta}(s) \\ \theta_{\text{tilt}}(s) \\ \theta_{\text{yaw}}(s) \end{bmatrix} = \begin{bmatrix} K_{\text{cpc}}(s) & 0 \\ 0 & K_{\text{ipc}}(s) \end{bmatrix} \begin{bmatrix} \omega(s) \\ M_{\text{tilt}}(s) \\ M_{\text{yaw}}(s) \end{bmatrix}, \tag{5.33}$$

where $K_{\text{cpc}}(s) \in \mathcal{R}$ is the CPC and $K_{\text{ipc}}(s) \in \mathcal{R}^{2\times 2}$ is the IPC, that are adopted from Chap. 4. Notice that in Fig. 5.3, the tower controller $K_{\text{t}} \in \mathcal{R}$ introduces an additional collective pitch signal:

$$\bar{\theta}(s) = K_{\text{cpc}}(s)\omega(s) + K_{\text{t}}(s)\hat{\dot{X}}_{\text{fa}}(s), \tag{5.34}$$

where the estimate of the fore-aft tower velocity $\hat{\dot{X}}_{\text{fa}}$ is obtained from the estimator $\Gamma_{\text{ob}} \in \mathcal{R}^{1\times(n_u+n_y)}$ according to:

$$\hat{\dot{X}}_{\text{fa}}(s) = \Gamma_{\text{ob}}(s) \begin{bmatrix} u_{\text{cm}}(s) \\ y_{\text{cm}}(s) \end{bmatrix}, \tag{5.35a}$$

$$= \left[\Gamma_{\text{ob}}^{(1,1)}(s)\ \Gamma_{\text{ob}}^{(1,2:6)}(s)\right] \begin{bmatrix} u_{\text{cm}}(s) \\ y_{\text{cm}}(s) \end{bmatrix}. \tag{5.35b}$$

Substituting (5.35) into (5.34) yields:

$$\bar{\theta}(s) = K_{\text{mod}}(s)\omega(s) + F(s)\left[\theta_{\text{tilt}}(s)\ \theta_{\text{yaw}}(s)\ y_{\text{cm}}(s)\right]^T, \tag{5.36a}$$

where $K_{\text{mod}} \in \mathcal{R}$ and $F \in \mathcal{R}^{1\times 5}$ are:

$$K_{\text{mod}}(s) = \frac{K_{\text{cpc}}(s)}{1 - K_{\text{t}}(s)\Gamma_{\text{ob}}^{(1,1)}(s)},\ F(s) = \frac{K_{\text{t}}(s)\Gamma_{\text{ob}}^{(1,2:6)}(s)}{1 - K_{\text{t}}(s)\Gamma_{\text{ob}}^{(1,1)}(s)}. \tag{5.36b}$$

Thus the observer introduces undesirable, but inevitable coupling from the tower controller to the CPC. With a view towards suppressing this, the tower controller is designed as an inverse notch filter with gain concentrated at the tower resonant frequency, which is above the bandwidth of the CPC:

$$K_{\text{t}}(s) := K_{\text{p}} \left(\frac{s^2 + 2D_1\omega_{\text{t}}s + \omega_{\text{t}}^2}{s^2 + 2D_2\omega_{\text{t}}s + \omega_{\text{t}}^2} \right). \tag{5.37}$$

The parameters for the tower controller K_{t} are $K_{\text{p}} = 0.03$, $D_1 = 5$ and $D_2 = 0.01$, and the magnitude bode diagram of this tower controller is shown in Fig. 5.4.

Figure 5.5 depicts the loop gain of the original CPC loop $G_{\text{cpc}}(s)K_{\text{cpc}}(s)$ and the modified loop $G_{\text{cpc}}(s)K_{\text{mod}}(s)$ and shows that the gain of the original loop $G_{\text{cpc}}(s)K_{\text{cpc}}(s)$ below the crossover frequency remains unaffected by the additional tower controller. Consequently, the gain and phase margins are largely unaffected, indicating little influence of the tower controller upon the CPC. In addition, the sensitivity plots in Fig. 5.6 shows that the additional couplings caused by the tower controller, mapping θ_{tilt}, θ_{yaw}, y_{cm} to $\bar{\theta}$ are insignificant.

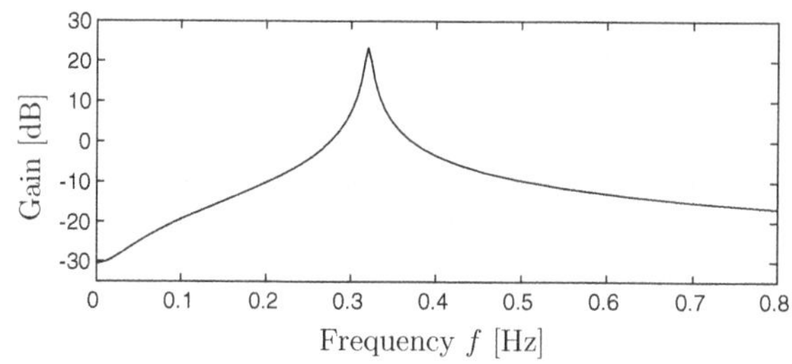

Fig. 5.4 Magnitude bode plot of the tower controller $K_t(s)$

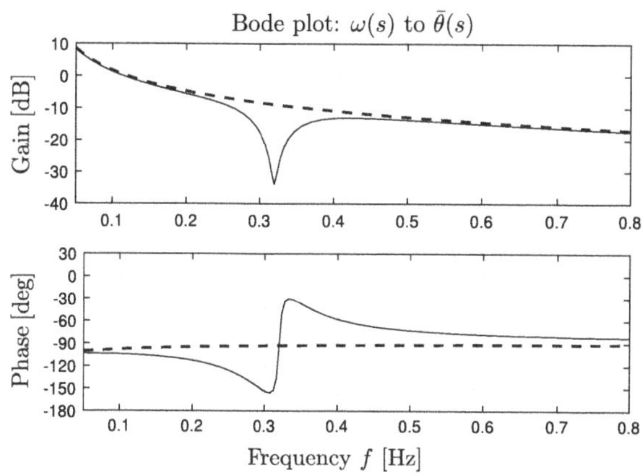

Fig. 5.5 Bode plot of the original CPC loop $G_{\mathrm{cpc}}(s)K_{\mathrm{cpc}}(s)$ (dashed line) and the modified loop $G_{\mathrm{cpc}}(s)K_{\mathrm{mod}}(s)$ by the additional tower controller (solid line)

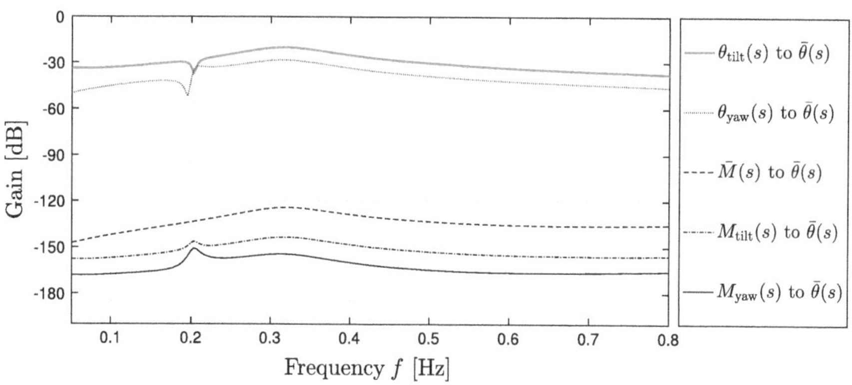

Fig. 5.6 The magnitudes of the transfer function $F(s)$ mapping θ_{tilt}, θ_{yaw}, $\bar{M}$, M_{tilt}, M_{yaw} to $\bar{\theta}$

5.7 Numerical Results and Discussion

This section presents simulation results to demonstrate the performance of the proposed estimator and estimation-based controller for the tower fore-aft motion. The turbine model employed in this work is the NREL 5MW turbine (Jonkman, Butterfield, Musial, & Scott, 2009) and the simulations are carried out on FAST (Jonkman & Buhl Jr., 2005). This turbine model is of much greater complexity than the linear model (5.31). All degrees-of-freedom were enabled, including flap-wise and edge-wise blade modes, in addition to the tower and shaft dynamics. Simulations in this study were conducted under a turbulent wind field with a mean wind speed 18 ms^{-1} (all model parameters were linearised around this operating point) chosen since this value is near the centre of the range of wind speeds covering the above-rated wind conditions. A turbulence intensity of 16% was used and was generated from TurbSim (Jonkman, 2009).

5.7.1 *Estimator Performance*

The performance was compared by examining the estimated signals of the tower fore-aft velocity and the actual measurements from the simulation turbine. The time histories of both the estimate alongside actual signal of the tower-top fore-aft velocity are shown in Fig. 5.7. It can be seen that good agreement is achieved between the estimated and actual signals, with a steady-state variance in the estimation error of order 10^{-3} m^2s^{-2}, as shown in Fig. 5.8.

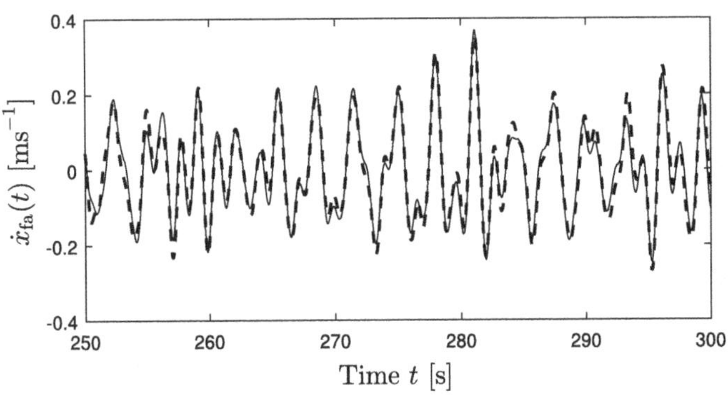

Fig. 5.7 Estimate $\hat{\dot{x}}_{fa}(t)$ (dashed line) and actual measurement $\dot{x}_{fa}(t)$ (solid line) of the tower-top velocity signal

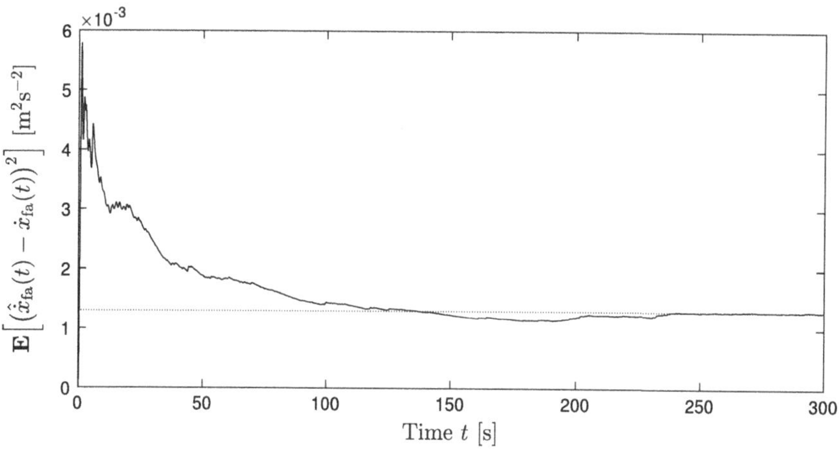

Fig. 5.8 Error variance (solid line) converges to a steady-state value (dotted line)

5.7.2 Controller Performance

To demonstrate the use of the estimated tower signals for control purposes, two controllers were compared: a baseline controller consisting of the CPC and IPC, from (Lio et al., 2015) and the proposed control architecture including the tower controller (5.37) employing the tower motion estimate. As shown in Fig. 5.9a, b, the tower vibrations were suppressed effectively by the proposed tower controller, with a marginal associated increase in blade pitch activity. Specifically, the standard deviation of the tower motion was 18% less with the tower controller for a 9% increase in the standard deviation of the blade pitch angle rates.

Of particular interest is the impact upon the fore-aft tower bottom moment M_{tb}, since this is a typical indicator of tower fatigue loads (Burton et al., 2011). The time history of this is shown in Fig. 5.9c, where significant reductions are achieved by the additional tower damping controller. Fig. 5.9d displays the load spectrum of the tower bottom moments and reveals a significant 87% load attenuation on M_{tb} at the tower resonant frequency (0.32 Hz), as was to be expected from the use of the inverse notch filter in the tower controller. In addition, the loads at frequencies below the blade rotational frequency (0.2 Hz) remained mostly unchanged, again indicating minimal coupling between the CPC tower controller. The results from the simulation are summarised in Table 5.2.

5.8 Summary

This chapter was concerned with the problem of attempting to estimate tower motion from blade-load measurements alone, in order to reduce sensor count. The coupling between states in both rotating and fixed frames of reference led to an initial system

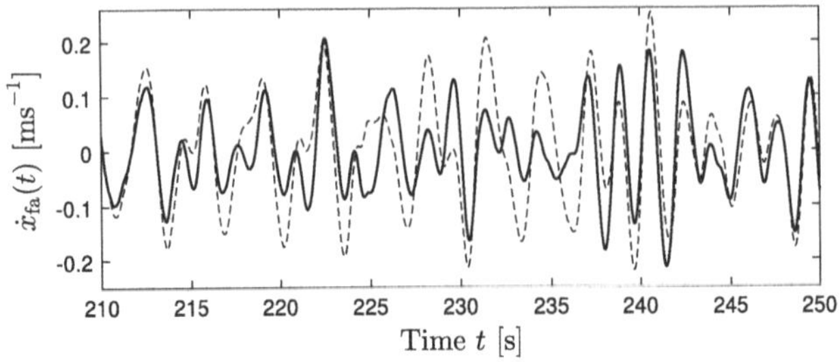

(A) Time history of the fore-aft tower-top velocity.

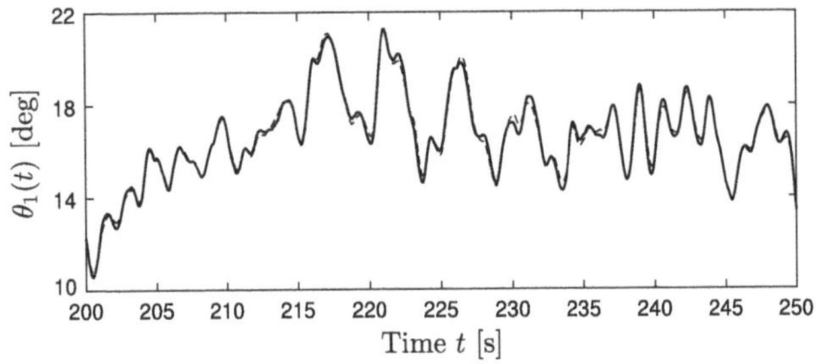

(B) Time history of the pitch activities of blade 1. Similar behaviours are observed for the remaining blades.

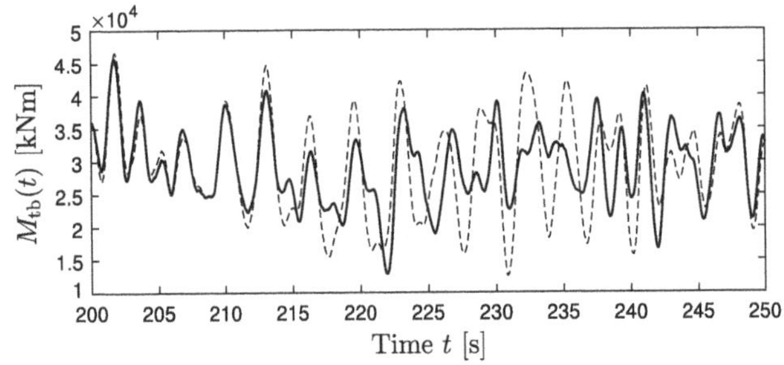

(C) Time history of the tower bottom moment.

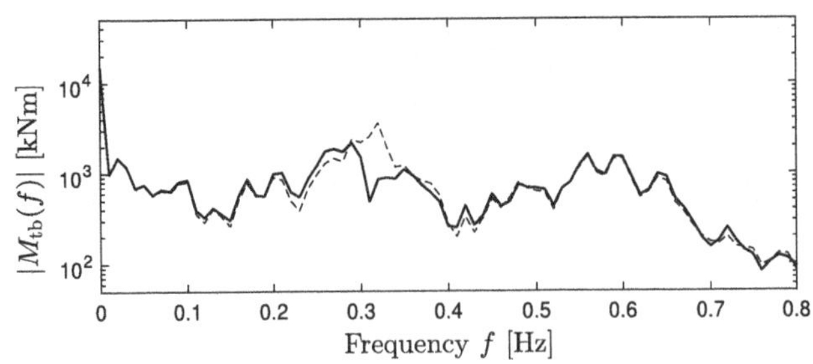

(D) Frequency spectrum of the tower bottom moments.

Fig. 5.9 Performance of the baseline controller (- -) compared to the proposed tower controller (–)

Table 5.2 Controller performance comparisons

	Baseline controller	Proposed controller
std($\dot{x}_{\text{fa}}$) (ms^{-1})	0.087 (100%)	0.072 (82%)
std($\dot{\theta}_1$) (deg s^{-1})	2.75 (100%)	3.00 (109%)
$\lvert M_{\text{tb}} \rvert$ at ω_{t} (MNm)	3.60 (100%)	0.48 (13%)

Note that std denotes the standard deviation. The percentage in brackets represents the relative difference to the baseline controller

model that was linear but time-varying, and so the key contribution of this chapter lay in simplifying this to a LTI model. This was achieved by employing the Coleman Transforms, and including the frequency-splitting effects of these in the resulting time-invariant model. Having verified observability, a state estimator was synthesised that produced good reconstructions of the tower fore-aft motion, based solely upon the blade-load measurements. This was subsequently used in a tower damping controller which was included in an architecture that augmented, rather than replaced the existing CPC and IPC, and was designed in such a way as to ensure negligible coupling between these controllers. Closed-loop simulations upon a high fidelity and nonlinear turbine model showed the estimation-based tower controller achieve an 87% reduction in the magnitude of the tower-base bending moment at the tower resonant frequency, for a modest 9% increase in the standard deviation of the blade pitch rates.

So far, the topics covered by Chaps. 4 and 5 were mainly focused on the feedback aspect of the IPC. In the coming chapters, the uses of upcoming wind information are investigated and their benefits for improving load reductions are demonstrated. The problem of formulating an MPC layer upon an existing feedback controller is examined in the following chapter.

References

Bittanti, S., & Colaneri, P. (2009). *Periodic systems*. London: Springer.

Bossanyi, E. A. (2000). The Design of closed loop controllers for wind turbines. *Wind Energy*, *3*(3), 149–163.

Bossanyi, E. A. (2003a). Individual blade pitch control for load reduction. *Wind Energy*, *6*(2), 119–128.

Bossanyi, E. A. (2003b). Wind turbine control for load reduction. *Wind Energy*, *6*(3), 229–244.

Brockett, R. W. (1970). *Finite dimensional linear systems*. New York: Wiley.

Burton, T., Jenkins, N., Sharpe, D., & Bossanyi, E. (2011). *Wind energy handbook*. Chichester, UK: Wiley.

Chen, C. T. (1984). *Linear system theory and design*. Oxford: Oxford University Press.

D'Angelo, H. (1970). *Linear time-varying systems: Analysis and synthesis*. Boston: Allyn and Bacon.

Dunne, F., Pao, L. Y., Wright, A. D., Jonkman, B., & Kelley, N. (2011). Adding feedforward blade pitch control to standard feedback controllers for load mitigation in wind turbines. *Mechatronics*, *21*, 682–690.

Evans, M. A., Cannon, M., & Kouvaritakis, B. (2015). Robust MPC tower damping for variable speed wind turbines. *IEEE Transactions on Control Systems Technology*, *23*(1), 290–296.

Feng, X., Patton, R., & Wang, Z. (2014). Sensor fault tolerant control of a wind turbine via Takagi-Sugeno fuzzy observer and model predictive control. In *2014 UKACC International Conference on Control (CONTROL)* (pp. 480–485). IEEE.

Johnson, C. D. (1971). Accommodation of external disturbances in linear regulator and servomechanism problems. *IEEE Transactions on Automatic Control*, *16*(6), 635–644.

Jonkman, B. (2009). TurbSim User's Guide TurbSim User's Guide (Technical Report). National Renewable Energy Laboratory (NREL).

Jonkman, J., & Buhl Jr, M. (2005). *FAST User's Guide* (Technical Report). National Renewable Energy Laboratory (NREL).

Jonkman, J., Butterfield, S., Musial, W., & Scott, G. (2009). *Definition of a 5-MW Reference Wind Turbine for Offshore System Development* (Technical Report). Golden, CO: National Renewable Energy Laboratory (NREL).

Kalman, R. (1959). On the general theory of control systems. *IRE Transactions on Automatic Control*, *4*(3), 110–110.

Kalman, R. E. (1960). A new approach to linear filtering and prediction problems. *Journal of Basic Engineering*, *82*(1), 35.

Laks, J., Pao, L., Wright, A., Kelley, N., & Jonkman, B. (2011). The use of preview wind measurements for blade pitch control. *Mechatronics*, *21*(4), 668–681.

Larsen, T. J., & Hanson, T. D. (2007). A method to avoid negative damped low frequent tower vibrations for a oating, pitch controlled wind turbine. In *Proceedings of the Science of Making Torque from Wind*.

Leithead, W., Dominguez, S., & Spruce, C. (2004). Analysis of Tower/Blade interaction in the cancellation of the tower fore-aft mode via control. *EWEC*.

Leithead, W., Neilson, V., & Dominguez, S. (2009). Alleviation of unbalanced rotor loads by single blade controllers. In *European Wind Energy Conference*.

Leithead, W., Neilson, V., Dominguez, S., & Dutka, A. (2009). A novel approach to structural load control using intelligent actuators. In *2009 17th Mediterranean Conference on Control and Automation* (pp. 1257–1262). IEEE.

Lio, W. H., Jones, B. L., Lu, Q., & Rossiter, J. A. (2015). Fundamental performance similarities between individual pitch control strategies for wind turbines. *International Journal of Control*, 1–16.

Lu, Q., Bowyer, R., & Jones, B. (2015). Analysis and design of Coleman transformbased individual pitch controllers for wind-turbine load reduction. *Wind Energy*, *18*(8), 1451–1468.

Markou, H., Buhl, T., Marrant, B., & van Engelen, T. (2006). Morphological Study of Aeroelastic Control Concepts for Wind Turbines. *ECN Report*.

Montagnier, P., Spiteri, R. J., & Angeles, J. (2004). The control of linear time-periodic systems using Floquet-Lyapunov theory. *International Journal of Control*.

Odgaard, P. F., & Stoustrup, J. (2013). Fault tolerant control of wind turbines: A benchmark model. *IEEE Transactions on Control Systems Technology*, *21*(4), 1168–1182.

Pao, L., & Johnson, K. (2009). A tutorial on the dynamics and control of wind turbines and wind farms. In *Proceedings of ACC*.

Sami, M., & Patton, R. J. (2012). *Global wind turbine FTC via T-S fuzzy modelling and control* (Vol. 45, No. 20). IFAC.

Selvam, K., Kanev, S., van Wingerden, J. W., van Engelen, T., & Verhaegen, M. (2009). Feedback-feedforward individual pitch control for wind turbine load reduction. *International Journal of Robust and Nonlinear Control*, *19*(1), 72–91.

Spruce, C. J., & Turner, J. K. (2013). Tower vibration control of active stall wind turbines. *IEEE Transactions on Control Systems Technology*, *21*(4), 1049–1066.

van Engelen, T. G. (2006). Design model and load reduction assessment for multirotational mode individual pitch control (higher harmonics control). In *Proceedings of European Wind Energy Conference*.

Chapter 6
Feed-Forward Model Predictive Control Design Based upon a Feedback Controller

6.1 Overview and Contribution

The potential use of upcoming measurement has motivated research on preview control to improve set-point tracking and disturbance rejection beyond that achievable via conventional feedback control. Such preview controllers, typically based upon model predictive control (MPC) for its constraint handling properties, could potentially introduce an additional feedback loop that therefore alters the closed-loop dynamics of the existing feedback compensator. This can result in a deterioration of the nominal robustness properties[1] and performance of the existing closed-loop. Therefore, the aim of this chapter is to formulate a modular MPC layer on top of a given output-feedback controller, with a view to retaining the nominal closed-loop robustness and frequency-domain properties of the latter. And a key result is derived that proves that the proposed modular MPC layer for handling the advance knowledge impacts upon the existing closed-loop system if and only if constraint violations are expected. The separable nature of the proposed control structure enables clear and transparent quantification of the benefits gained by using preview control and constraint handling, beyond that of the underlying feedback compensator.

This chapter focuses on theoretical development of the MPC layer. The discussions of the application to wind turbines are presented in Chap. 7. This chapter is setting up some important ground work for the coming chapter.

The remainder of this chapter is structured as follows. Section 6.2 presents an introduction and motivation of the chapter. Preliminaries on modelling of the closed-loop system are presented in Sect. 6.3. Section 6.4 demonstrates the formulation of the proposed feed-forward MPC layer upon the existing output-feedback controller, which results in a simple but systematic form. In Sect. 6.5, the key results of the

[1] The robustness property here implies the robust stability margin, that is a generalisation of the gain and phase margins for multi-input-multi-output systems.

W. H. (Alan) Lio, *Blade-Pitch Control for Wind Turbine Load Reductions*,
Springer Theses, https://doi.org/10.1007/978-3-319-75532-8_6

conditions are derived that prevent the MPC layer introducing an additional feedback loop to the closed-loop dynamics and discussions on robust stability and tuning of the proposed MPC layer are also presented. It is followed by numerical examples in Sect. 6.6 and conclusion in Sect. 6.7.

6.2 Introduction

In many control applications, preview knowledge is available for improving tracking quality and disturbance rejection. Model predictive control (MPC) has been used by an increasing number of applications to utilise the preview knowledge because, in principle, the information is incorporated in a systematic fashion. However, typically, robustness and frequency domain properties are not well considered in a standard MPC design and indeed the optimisation implies coupling between feedforward and feedback information. To address this weakness, this chapter aims to decouple the feedback and feed-forward design of the MPC problem such that the feedback control law focusses on the closed-loop properties such as sensitivity, whilst the feed-forward input is purely based on advance knowledge without any corrections linked to the measurements from the plant.

Much research in recent years has focused on how to incorporate the full preview knowledge into an MPC feedback design in a more systematic and optimal fashion. The earlier works by Rossiter and Grinnell (1995, 1996) and demonstrated very clearly that the transient behaviour during set-point changes can be improved much further by adopting a more systematic separate design of the feed-forward compensator on a finite-horizon MPC design (Clarke, Mohtadi, & Tus, 1987). Later, for tracking a sinusoid trajectory, a study by Wang and Rossiter (2008) developed an adaptive algorithm to optimise an MPC feed-forward compensator, solely relying upon the advance measurements. Similarly, a study by Rossiter and Valencia-Palomo (2009) also proposed a two stage design, where the feed-forward compensator is optimised based on the given set-point trajectory, separating from the feedback MPC design. Recently, in Valencia-Palomo, Rossiter, and López-Estrada (2014), the two stage design (Rossiter & Valencia-Palomo, 2009) was successfully implemented on hardware with low processing power. Furthermore, studies by Carrasco and Goodwin (2011a, 2011b); Goodwin, Carrasco, and Mayne (2011) proposed a feed-forward MPC design, aiming to utilise the feed-forward control law for tracking and disturbance rejection performance, whilst the feedback controller mainly for robustifying the closed-loop. Notice that the nature of these works (Carrasco & Goodwin, 2011a, 2011b; Goodwin et al., 2011) slightly differs from those seeking to systematically incorporate the feed-forward knowledge to the feedback MPC design.

A large part of the feed-forward MPC studies assumed that the design was based on a state-feedback controller. Nonetheless, for some applications, output-feedback controller is often employed, which is synthesised using frequency-domain techniques for satisfying some robustness and performance specifications given in the frequency-domain. As a consequence, this begs the question: 'Is there a systematic way to incorporate the merits of the MPC such as the capability of handling constraint and preview knowledge into an existing output-feedback controller?'. This problem may seem immediately obvious to be addressed simply by a standard MPC formulating around the underlying closed-loop dynamics. However, given that the constraint handling features of the feed-forward MPC depend upon the predictions of the closed-loop dynamics, optimising such predictions could indeed potentially introduce an additional feedback loop to the original closed-loop, that would deteriorate the properties of the existing controller.

This chapter proposes an MPC layer based on a known robust output-feedback controller where the proposed MPC layer handles feed-forward information and constraints whilst retaining the nominal robustness and performance of the underlying feedback controller. A further key focus of this chapter is to investigate under what conditions the outer control layer design is separated from impacting upon the original closed-loop nominal stability and robustness. The separate nature of this MPC layer offers immediate benefits from an industry perspective, since the feed-forward control with constraint handling capability can be implemented without replacing the existing feedback controller. Furthermore, the separate controller structure provides a clear framework to quantify the benefits of the use of advance knowledge over the nominal output-feedback strategy, as well as the impact of constraint handling.

For the sake of brevity and clarity, problems regarding steady-state due to the active constraint (see. e.g. Dughman & Rossiter, 2015; Limon, Alvarado, Alamo, & Camacho, 2008; Rawlings, Bonné, Jørgensen, Venkat, & Jørgensen, 2008; Shead, Muske, & Rossiter, 2010) are not particularly considered in this chapter since the purpose of this chapter is to present a simple feed-forward MPC formulation upon an output-feedback controller.

6.3 Modelling

This section introduces basic definitions of the plant and the chosen robust feedback controller.

6.3.1 Plant Model

Assume the linear model of the plant $G(s) \in \mathcal{R}^{n_y \times n_u}$ and the disturbance model $G_\mathrm{d}(s) \in \mathcal{R}^{n_y \times n_d}$ are strictly proper functions, which can be described in discrete-time state-space forms as follows:

$$\begin{aligned} x_{\mathrm{p}_{k+1}} &= A_\mathrm{p} x_{\mathrm{p}_k} + B_\mathrm{p} u_k + B_{\mathrm{d}_\mathrm{p}} d_k, \\ y_k &= C_\mathrm{p} x_{\mathrm{p}_k}, \end{aligned} \tag{6.1}$$

where $u_k \in \mathbb{R}^{n_u}$, $y_k \in \mathbb{R}^{n_y}$, $d_k \in \mathbb{R}^{n_d}$ and $x_{\mathrm{p}_k} \in \mathbb{R}^{n_{x\mathrm{p}}}$ represent the input, output, disturbance and state of the plant, respectively, whilst the subscript p denotes the plant. The plant model (6.1) is assumed to be controllable and observable.

6.3.2 Feedback Controller

Let the robust nominal output-feedback controller stabilising the linear model (6.1) be given by $K(s) \in \mathcal{R}^{n_u \times n_y}$, with its discrete-time state-space form is as follows:

$$\begin{aligned} x_{\kappa_{k+1}} &= A_\kappa x_{\kappa_k} - B_\kappa y_k, \\ u_k &= C_\kappa x_{\kappa_k} - D_\kappa y_k, \end{aligned} \tag{6.2}$$

where the vector $x_{\kappa_k} \in \mathbb{R}^{n_{x_\kappa}}$ represents the state of the controller and the subscript κ denotes controller. The controller model (6.2) is assumed to be controllable and observable.

Remark 6.1 The purpose of this chapter is to incorporate the constraints handling capability and advance knowledge into an existing output-feedback controller that satisfies the design specifications provided in the frequency-domain. For example, load attenuations at a specific frequency range or Nyquist stability criterion. Thus, the feedback controller (6.2) is assumed to be synthesised by frequency-domain design methods, such as robust $\mathcal{H}_\infty$ loop-shaping method, for example, the baseline IPC controller in Sect. 2.8 in Chap. 2.

6.3.3 Closed-Loop Model

Consequently, by combining the linear model (6.1) and controller (6.2), the closed-loop dynamic system model can be expressed in the following form:

$$\begin{bmatrix} x_{\mathrm{p}k+1} \\ x_{\kappa k+1} \\ u_k \end{bmatrix} = \underbrace{\begin{bmatrix} A_\mathrm{p} & 0 & B_\mathrm{p} \\ -B_\kappa C_\mathrm{p} & A_\kappa & 0 \\ 0 & 0 & I \end{bmatrix}}_{A} \underbrace{\begin{bmatrix} x_{\mathrm{p}k} \\ x_{\kappa k} \\ u_{k-1} \end{bmatrix}}_{x_k} + \underbrace{\begin{bmatrix} B_\mathrm{p} \\ 0 \\ I \end{bmatrix}}_{B} \Delta u_k + \underbrace{\begin{bmatrix} B_{\mathrm{d}_\mathrm{p}} \\ 0 \\ 0 \end{bmatrix}}_{B_\mathrm{d}} d_k, \tag{6.3a}$$

$$\Delta u_k = \begin{bmatrix} -D_\kappa C_\mathrm{p} & C_\kappa & -I \end{bmatrix} x_k = K x_k, \tag{6.3b}$$

$$y_k = \begin{bmatrix} C_\mathrm{p} & 0 & 0 \end{bmatrix} x_k = C x_k. \tag{6.3c}$$

The pair $\{A, B\}$ is controllable, whilst the pair $\{A, [C, K]^T\}$ is observable. Notice that an incremental input form of the closed-loop model is employed for the simplicity reasons of implementing input rate constraints in the MPC layer.

6.3.4 Steady-State Target Calculator

Assume the upcoming disturbance is available, a typical steady-state target calculator can be used to generate the desired input trajectory with respect to the desired output trajectory. Consider the deviation of the state $\tilde{x}_k = x_k - x_k^{\mathrm{s}}$, input $\Delta\tilde{u}_k = \Delta u_k - \Delta u_k^{\mathrm{s}}$ and output $\tilde{y} = y_k - r_k$ from the steady-state $x_k^{\mathrm{s}}, \Delta u_k^{\mathrm{s}}, r_k^{\mathrm{s}}$, then the closed-loop mode (6.3a) can be expressed in terms of the deviation as follows:

$$\tilde{x}_{k+1} = A\tilde{x}_k + B\Delta\tilde{u}_k, \tag{6.4a}$$

$$\Delta\tilde{u}_k = K\tilde{x}_k, \tag{6.4b}$$

$$\tilde{y}_k = C\tilde{x}_k. \tag{6.4c}$$

Assuming the set-point r_k^{s} is zero since this chapter considers a regulator problem, the steady-state $\Delta u_k^{\mathrm{s}}, x_k^{\mathrm{s}}$ can be calculated from a typical steady-state target calculator (e.g. Muske & Rawlings, 1993), defined as follows:

$$\begin{bmatrix} x_k^{\mathrm{s}} \\ \Delta u_k^{\mathrm{s}} \end{bmatrix} = \begin{bmatrix} I - A & -B \\ C & 0 \end{bmatrix}^{-1} \begin{bmatrix} B_{\mathrm{b}} \\ 0 \end{bmatrix} d_k, \tag{6.5a}$$

$$= \begin{bmatrix} K_{xd} \\ K_{ud} \end{bmatrix} d_k. \tag{6.5b}$$

Thus, if the upcoming disturbance d_k is available, the control law (6.4b) becomes of the following form:

$$\Delta u_k = K x_k + \underbrace{(K_{ud} - K K_{xd}) d_k}_{\Delta u_k^{\mathrm{ff}}}, \tag{6.6}$$

Notice that the term $(K_{ud} - K K_{xd})$ is a feed-forward gain and Δu_k^{ff} denotes the feed-forward input to the original feedback control law (6.2).

Remark 6.2 If the controller (6.2) contains an integrator, the matrix $\begin{bmatrix} I - A & -B \\ C & 0 \end{bmatrix}$ could be non-invertible. A simple workaround is to place the pole of the integrator of (6.2) slightly off from the origin.

The feed-forward structure (6.6) only handles steady-state offset effectively. Inspection of the control law (6.6) reveals that the feed-forward input Δu_k^{ff} only takes into account the upcoming disturbance at k. Thus, the entire upcoming trajectory are not fully utilised.

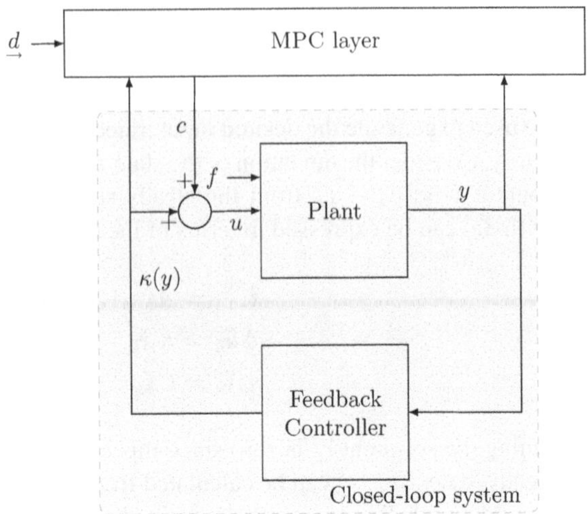

Fig. 6.1 Schematic of model predictive control layer on top of an existing feedback controller. Based upon the controller input $\kappa(y)$, plant output y, the layer optimises the perturbation c that can handle constraints and act upon feed-forward information $\underset{\rightarrow}{d}$. The notation f denotes the disturbance to the plant

6.4 Design of the MPC Layer

This section describes the design of the MPC layer to compliment the output-feedback controller (6.2) in the previous section. The architecture combining the proposed control layer and the separate feedback controller is shown in Fig. 6.1, where u, y and f denote the input, output and actual disturbance of the plant, respectively. Based on the control action of the existing controller $\kappa(y)$ and measured disturbance $\underset{\rightarrow}{d}$, the proposed MPC layer optimises a perturbation c to handle constraints as well as act upon the feed-forward information.

6.4.1 Augmentation of Perturbation Into the Underlying Control Law

The MPC layer formulation in this chapter adopts a dual-mode closed-loop paradigm (e.g. Rossiter, 2003; Rossiter, Kouvaritakis, & Rice, 1998), whereby the degrees-of-freedom (d.o.f) $c_k \in \mathbb{R}^{n_u}$ within the predictions are defined around a stabilising feedback control law such that the input can be parameterised as follows:

$$\tilde{u}_{i|k} = \begin{cases} K\tilde{x}_{i|k} + c_{i|k}, & \forall i = \{0, \cdots, n_c - 1\}, \\ K\tilde{x}_{i|k}, & \forall i \geq n_c, \end{cases} \tag{6.7}$$

The premise behind this approach is that the MPC perturbation $c_k \neq 0$ if and only if constraints are active and feed-forward knowledge is available; obviously when $c_k = 0$ the underlying feedback controller operates unaffected. Such a feature is particularly useful in formulating an MPC layer on top of a given feedback controller. Notice that the predicted perturbation sequence $\underset{\rightarrow k}{c} = [c_{0|k}, c_{1|k}, \ldots, c_{n_c-1|k}]^T \in \mathbb{R}^{n_u n_c}$ is optimised over the control horizon n_c, whilst beyond n_c, the closed-loop dynamics are governed solely by the given feedback control law.

6.4.2 Formulation of Cost Function

Let the predicted cost function and associated predictions employed in the MPC layer take the following form:

$$J_k = \sum_{i=0}^{\infty} \tilde{x}_{i|k}^T Q \tilde{x}_{i|k} + \Delta\tilde{u}_{i|k}^T R \Delta\tilde{u}_{i|k}, \tag{6.8a}$$

$$\text{s.t.} \quad \tilde{x}_{i+1|k} = A\tilde{x}_{i|k} + B\Delta\tilde{u}_{i|k}, \quad \forall i \geq 0, \tag{6.8b}$$

$$\tilde{u}_{i|k} = \begin{cases} K\tilde{x}_{i|k} + c_{i|k}, & \forall i = \{0, \ldots, n_c - 1\}, \\ K\tilde{x}_{i|k}, & \forall i \geq n_c, \end{cases} \tag{6.8c}$$

$$d_{i|k} = \begin{cases} d_{k+i} & \forall i = \{0, \ldots, n_a - 1\}, \\ d_{k+n_a-1} & \forall i \geq n_a \end{cases} \tag{6.8d}$$

$$u_{i|k}^s = K_{ud} d_{i|k}, \; x_{i|k}^s = K_{xd} d_{i|k}. \tag{6.8e}$$

where $Q \in \mathbb{R}^{n_x \times n_x}$ and $R \in \mathbb{R}^{n_u \times n_u}$ denote the weighting matrices that specify the penalties on state and input, respectively. The predictions of state, input, disturbance and steady-state are denoted by (6.8b), (6.8c), (6.8d) and (6.8e), respectively. For brevity, disturbance d_k is assumed to be known perfectly in advance and beyond the preview horizon n_a, the predicted disturbance is implicitly assumed to be constant thereafter $d_{i|k} = d_{k+n_a-1}$, $\forall i \geq n_a$. If the measurements of the upcoming disturbance cannot be obtained accurately, an observer design can be employed to estimate the unknown disturbance (Pannocchia & Rawlings, 2003).

Remark 6.3 The predictions of the disturbance (6.8d) is essential to ensure that the full upcoming knowledge over the preview horizon n_a is taken into account within the cost function (6.8a).

The predictions of state (6.8b), input (6.8c), disturbance (6.8d) and steady-state (6.8e) can be expressed in a more convenient and compact autonomous model form, where its state $z_{i|k} \in \mathbb{R}^{n_z}$ consists of the state $x_{i|k}$ of the model, perturbations $\underset{\rightarrow k}{c}$ and future disturbance $\underset{\rightarrow k}{d} = [d_{0|k}, d_{1|k}, \ldots, d_{n_a-1|k}]^T \in \mathbb{R}^{n_d n_a}$, defined as follows:

$$z_{i+1|k} = \Psi z_{i|k}, \tag{6.9a}$$

where the initial state $z_{0|k} = [x_{0|k}, \underset{\rightarrow k}{c}, \underset{\rightarrow k}{d}]^T$ and Ψ is defined as:

$$\Psi = \begin{bmatrix} \Phi & BE & (I - \Phi)K_{xd}E \\ 0 & M_c & 0 \\ 0 & 0 & M_d \end{bmatrix}, \tag{6.9b}$$

$$E \underset{\rightarrow k}{c} = c_{0|k}, \quad E \underset{\rightarrow k}{d} = d_{0|k}, \tag{6.9c}$$

$$M_c \underset{\rightarrow k}{c}^T = [c_{1|k}, \ldots, c_{n_c-1|k}, 0]^T, \tag{6.9d}$$

$$M_d \underset{\rightarrow k}{d}^T = [d_{1|k}, \ldots, d_{n_a-1|k}, d_{n_a-1|k}]^T. \tag{6.9e}$$

where $\Phi = A + BK \in \mathbb{R}^{n_x \times n_x}$ is Hurwitz and the shift matrices $M_c \in \mathbb{R}^{n_u n_c \times n_u n_c}$ and $M_d \in \mathbb{R}^{n_d n_a \times n_d n_a}$ are defined as follows:

$$M_c = \begin{bmatrix} 0 & I & 0 & \cdots & 0 \\ 0 & 0 & I & \cdots & 0 \\ \vdots & \vdots & \vdots & \ddots & \vdots \\ 0 & 0 & 0 & \cdots & I \\ 0 & 0 & 0 & \cdots & 0 \end{bmatrix}, \quad M_d = \begin{bmatrix} 0 & I & 0 & \cdots & 0 \\ 0 & 0 & I & \cdots & 0 \\ \vdots & \vdots & \vdots & \ddots & \vdots \\ 0 & 0 & 0 & \cdots & I \\ 0 & 0 & 0 & \cdots & I \end{bmatrix}, \tag{6.10a}$$

and $E \in \mathbb{R}^{n_x \times n_u n_c}$ are described as follows:

$$E = \begin{bmatrix} I & 0 & 0 & \cdots & 0 \end{bmatrix}. \tag{6.10b}$$

Subsequently, the optimisation problem of (6.8), using the autonomous form (6.9), is described as follows:

$$J_k = \sum_{i=0}^{\infty} z_{i|k}^T (\Gamma_x^T Q \Gamma_x + \Gamma_u^T R \Gamma_u) z_{i|k}, \tag{6.11a}$$

$$\text{s.t.} \quad z_{i+1|k} = \Psi z_{i|k}, \quad \forall i \geq 0, \tag{6.11b}$$

where $\Gamma_x = \begin{bmatrix} I & 0 & -K_{xd} \end{bmatrix} \in \mathbb{R}^{n_x \times n_z}$ and $\Gamma_u = \begin{bmatrix} K & E & -K K_{xd} \end{bmatrix} \in \mathbb{R}^{n_u \times n_z}$. The infinite-horizon cost function (6.11a) can be compacted into a finite-horizon form, using the Lyapunov equation $\Psi^T S \Psi = S - W$ and $z_{i|k} = \Psi^i z_{0|k}$:

$$J_k = z_{0|k}^T \underbrace{\sum_{i=0}^{\infty} \Psi^{i^T} \underbrace{\Gamma_x^T Q \Gamma_x + \Gamma_u^T R \Gamma_u}_{W} \Psi^i}_{S} z_{0|k}, \tag{6.12a}$$

$$= \begin{bmatrix} x_{0|k} \\ \underset{\rightarrow k}{c} \\ \underset{\rightarrow k}{d} \end{bmatrix}^T S \begin{bmatrix} x_{0|k} \\ \underset{\rightarrow k}{c} \\ \underset{\rightarrow k}{d} \end{bmatrix} = \begin{bmatrix} x_{0|k} \\ \underset{\rightarrow k}{c} \\ \underset{\rightarrow k}{d} \end{bmatrix}^T \begin{bmatrix} S_x & S_{xc} & S_{xd} \\ S_{xc}^T & S_c & S_{cd} \\ S_{xd}^T & S_{cd}^T & S_d \end{bmatrix} \begin{bmatrix} x_{0|k} \\ \underset{\rightarrow k}{c} \\ \underset{\rightarrow k}{d} \end{bmatrix}, \tag{6.12b}$$

$$= \underset{\rightarrow k}{c}^T S_c \underset{\rightarrow k}{c} + 2\underset{\rightarrow k}{c}^T S_{xc}^T x_{0|k} + 2\underset{\rightarrow k}{c}^T S_{cd} \underset{\rightarrow k}{d} + \epsilon, \tag{6.12c}$$

where ϵ denotes the terms that are independent of $\underset{\rightarrow}{c}$ and $x_{0|k} = x_k$.

6.4.3 Constraint Formulations in Terms of Perturbations

Let the system be subject to constraints of the form:

$$y_{\min} \leq y_{i|k} \leq y_{\max}, \qquad \forall i \geq 0, \tag{6.13a}$$

$$u_{\min} \leq u_{i|k} \leq u_{\max}, \qquad \forall i \geq 0, \tag{6.13b}$$

$$\Delta u_{\min} \leq \Delta u_{i|k} \leq \Delta u_{\max}, \qquad \forall i \geq 0, \tag{6.13c}$$

Given that the state and input is captured in terms of the autonomous form (6.9), thus, the inequalities (6.13) can be written as follows:

$$H z_{i|k} \leq h, \qquad \forall i \geq 0, \tag{6.14a}$$

where

$$H z_{i|k} = \begin{bmatrix} y_{i|k} & -y_{i|k} & u_{i|k} & -u_{i|k} & \Delta u_{i|k} & -\Delta u_{i|k} \end{bmatrix}^T, \tag{6.14b}$$

$$h = \begin{bmatrix} y_{\max} & -y_{\min} & u_{\max} & -u_{\min} & \Delta u_{\max} & -\Delta u_{\min} \end{bmatrix}^T. \tag{6.14c}$$

It is noted that to ensure no constraint violations, possible violations in (6.14) must be checked over an infinite prediction horizon, which would appear to be computationally impractical. However, it is well known (Gilbert & Tan, 1991) that there exists a sufficiently large horizon where any additional linear equalities of (6.14) for $i \geq n_\infty$ become redundant, assuming Φ is strictly Hurwitz, $c_{i|k} = 0$ for $i \geq n_c$, $d_{i|k}$ is bounded and the constraints contain the steady-state within their interior. Thus, the inequalities (6.14) can be expressed as a maximal controlled admissible set as following:

$$\begin{aligned} \mathcal{S} &= \{x_{i|k} \mid \exists \underset{\rightarrow}{c} : H\Psi^i z_{0|k} \leq h, \ \forall i = \{0, \cdots, n_\infty\}\}, \\ &= \{x_{i|k} \mid \exists \underset{\rightarrow}{c} : M x_k + N \underset{\rightarrow k}{c} + V \underset{\rightarrow k}{d} \leq b\} \end{aligned} \tag{6.15}$$

where the matrices (M, N, V and b) can be computed off-line by the admissible set algorithms (Gilbert & Tan, 1991; Pluymers, Rossiter, Suykens, & De Moor, n.d.).

Assumption 6.4.1 The disturbance $d_{i|k}$ is assumed to be bounded by a sensible range.

Assumption 6.4.2 The steady-state $y^s_{i|k}, u^s_{i|k}, \Delta u^s_{i|k}$ is assumed to be bounded by the constraint (6.13).

Assumption 6.4.1 is essential for the reasons of boundedness and convergence of the maximal controlled admissible set (6.15) and Assumption 6.4.2 indicates that the steady-state is assumed to be reachable. Thus, the problems regarding unreachable targets due to active constraints are not considered and nevertheless, these problems can be addressed by an on-line steady-state target calculator (e.g. Shead, Muske, & Rossiter, 2008) or soft-constraints (e.g. Zeilinger, Morari, & Jones, 2014). To sum up, the proposed MPC layer can be summarised by Algorithm 6.4.1.

Algorithm 6.4.1 (MPC layer) *At each sampling instant k:*

1. *Perform the optimisation*

$$\min_{\underset{\rightarrow k}{c}} \underset{\rightarrow k}{c}^T S_c \underset{\rightarrow k}{c} + 2\underset{\rightarrow k}{c}^T (S_{cd} \underset{\rightarrow k}{d} + S_{xc}^T x_{0|k}), \tag{6.16a}$$

$$\text{s.t.} \quad \mathcal{M} x_{0|k} + \mathcal{N} \underset{\rightarrow k}{c} + \mathcal{V} \underset{\rightarrow k}{d} \leq b. \tag{6.16b}$$

2. *Apply the first block element $c_k = c_{0|k}$ of the perturbation sequence $\underset{\rightarrow k}{c}$ within the embedded control law $\Delta\tilde{u}_k = K\tilde{x}_k + c_k$, where the perturbation sequence $\underset{\rightarrow k}{c}$ is the minimiser of the optimisation problem (6.16).*

6.5 Analysis of the MPC Layer

While the previous section presents insights on formulating an MPC layer on an existing output-feedback controller, this section demonstrates the conditions that ensures the nominal stability and robustness properties of the original closed-loop are retained.

6.5.1 *Main Results*

With the MPC layer and feed-forward input (6.6), the original input (6.3b) becomes as follows:

$$\Delta u_k = Kx_k + \Delta u_k^{\mathrm{ff}} + c_k. \tag{6.17}$$

The perturbation c_k handles the constraints and preview knowledge, however, this might not be true as illustrated in the following lemma.

Lemma 6.4 *The MPC layer introduces an additional feedback loop to the existing closed-loop system when constraints are inactive.*

Proof This can be demonstrated by an unconstrained minimisation of the cost (6.16a), where the perturbation is:

$$c_k = E\underset{\rightarrow k}{c} = -\underbrace{ES_c^{-1}S_{xc}^T}_{K_{cx}}x_k - \underbrace{ES_c^{-1}S_{cd}}_{K_{cd}}\underset{\rightarrow k}{d}, \tag{6.18}$$

Substituting (6.18) into (6.17) yields:

$$\Delta u_k = (K - K_{cx})x_k + \Delta u_k^{\mathrm{ff}} - K_{cd}\underset{\rightarrow k}{d}. \tag{6.19}$$

As the optimum c_k depends upon the state x_k, the underlying state feedback gain K is implicitly changing to $K - K_{cx}$. □

Lemma 6.4 indicates that the perturbation c_k from the MPC layer could potentially change the nominal stability and robustness properties of the underlying controller even when constraints are not active.

Lemma 6.5 *Suppose* $S_{xc}^T = 0$, *the cost function of* (6.16a) *is of the following form:*

$$J_k := \underset{\rightarrow k}{c}^T S_c \underset{\rightarrow k}{c} + 2\underset{\rightarrow k}{c}^T S_{cd}\underset{\rightarrow k}{d}, \tag{6.20}$$

then, $\underset{\rightarrow k}{c}$ *does not impact on the original closed-loop dynamics except when constraint violations are expected.*

Proof By performing an unconstrained minimisation of (6.20), the perturbation becomes:

$$c_k = -\underbrace{ES_c^{-1}S_{cd}}_{K_{cd}}\underset{\rightarrow k}{d}. \tag{6.21}$$

Now c_k does not depend upon $x_{0|k}$ and comprises solely a feed-forward input to the closed-loop system. Substituting (6.21) into (6.17) yields:

$$\Delta u_k = Kx_k + \Delta u_k^{\mathrm{ff}} - K_{cd}\underset{\rightarrow k}{d}. \tag{6.22}$$

Consequently, the nominal stability and robustness properties of the original closed-loop dynamics are not affected by the MPC layer. □

Corollary 6.5.1 To retain the nominal closed-loop dynamics, it is required that the perturbation c_k becomes independent of the feedback measurement x_k, in other words, as shown in Lemma 6.5, the term $\underset{\rightarrow k}{c}^T S_{xc}^T x_{0|k}$ in the cost function of Algorithm 6.4.1 needs to be zero.

Therefore, the main result of this chapter is as follows.

Theorem 6.6 *The unconstrained input perturbation sequence $\underset{\rightarrow}{c}$ from the additional MPC layer (Definition 6.4.1) has no impact on the original closed-loop dynamics if and only if $S_{xc}^T = 0$. For $S_{xc}^T = 0$, the cost function* (6.16a) *needs to embed some knowledge of the nominal output-feedback control law* (6.2) *such that the weights Q, R satisfy the following conditions:*

$$\Phi^T S_x \Phi - S_x + Q + K^T R K = 0, \tag{6.23a}$$

$$B^T S_x \Phi + R K = 0. \tag{6.23b}$$

Proof The proof is based on inspection of (6.12) and the Lyapunov equation $\Psi^T S \Psi = S - W$, which can be expressed as follows:

$$\begin{bmatrix} \Phi^T & 0 & 0 \\ E^T B^T & M_c^T & 0 \\ E^T B_d^T & 0 & M_d^T \end{bmatrix} \begin{bmatrix} S_x & S_{xc} & S_{xd} \\ S_{xc}^T & S_c & S_{cd} \\ S_{xd}^T & S_{cd}^T & S_d \end{bmatrix} \begin{bmatrix} \Phi & BE & B_d E \\ 0 & M_c & 0 \\ 0 & 0 & M_d \end{bmatrix} - \begin{bmatrix} S_x & S_{xc} & S_{xd} \\ S_{xc}^T & S_c & S_{cd} \\ S_{xd}^T & S_{cd}^T & S_d \end{bmatrix}$$
$$+ \begin{bmatrix} Q + K^T R K & K^T R E & -(Q + K^T R K) K_{xd} \\ E^T R K & E^T R E & -E^T R K K_{xd} \\ -K_{xd}^T (Q + K^T R K) & -K_{xd}^T K^T R E & K_{xd}^T (Q + K^T R K) K_{xd} \end{bmatrix} = 0. \tag{6.24}$$

To find the conditions where $S_{xc}^T = 0$, begin from the top-left equality of (6.24):

$$\Phi^T S_x \Phi - S_x + Q + K^T R K = 0, \tag{6.25}$$

which forms the first condition of Theorem 6.6. Note that this is the Lyapunov equation for the pre-determined feedback control law and since $x_{k+1} = \Phi x_k$ where the closed-loop dynamics Φ is asymptotically stable, given any $Q > 0$, there exist a unique $S_x > 0$ satisfying (6.25). Subsequently, considering the middle-left equality of (6.24):

$$E^T B^T S_x \Phi + M_c^T S_{xc}^T \Phi - S_{xc}^T + E^T R K = 0, \tag{6.26a}$$

and since $S_x > 0$, the condition for $S_{xc}^T = 0$ is if and only if :

$$E^T B^T S_x \Phi + E^T R K = 0, \tag{6.26b}$$

and (6.26b) can be further simplified:

$$B^T S_x \Phi + R K = 0. \tag{6.26c}$$

To show $B^T S_x \Phi + RK = 0$ is a necessary and sufficient condition for $S_{xc}^T = 0$, proofs of sufficiency and necessity are provided.

Proof of sufficiency: Suppose $B^T S_x \Phi + RK = 0$ and rewrite 6.26a as follows:

$$\begin{bmatrix} B^T S_x \Phi \\ 0 \\ \vdots \\ 0 \\ 0 \end{bmatrix} + \begin{bmatrix} 0 & 0 & \cdots & 0 & 0 \\ I & 0 & \cdots & 0 & 0 \\ 0 & I & \cdots & 0 & 0 \\ \vdots & \vdots & \ddots & \vdots & \vdots \\ 0 & 0 & \cdots & I & 0 \end{bmatrix} \underbrace{\begin{bmatrix} S_{xc_{\{1,:\}}}^T \Phi \\ S_{xc_{\{2,:\}}}^T \Phi \\ \ddots \\ S_{xc_{\{n_c-1,:\}}}^T \Phi \\ S_{xc_{\{n_c,:\}}}^T \Phi \end{bmatrix}}_{S_{xc}^T \Phi} - \underbrace{\begin{bmatrix} S_{xc_{\{1,:\}}}^T \\ S_{xc_{\{2,:\}}}^T \\ \ddots \\ S_{xc_{\{n_c-1,:\}}}^T \\ S_{xc_{\{n_c,:\}}}^T \end{bmatrix}}_{S_{xc}^T} + \begin{bmatrix} RK \\ 0 \\ \vdots \\ 0 \\ 0 \end{bmatrix} = 0, \tag{6.27a}$$

where $S_{xc}^T \in \mathbb{R}^{n_u n_c \times n_x}$ is expressed in terms of $S_{xc_{\{i,:\}}}^T \in \mathbb{R}^{n_u \times n_x}$ for $i \in \{1, \cdots, n_c\}$. Considering the first equality of (6.27a):

$$B^T S_x \Phi - S_{xc_{\{1,:\}}}^T + RK = 0. \tag{6.27b}$$

Substituting $B^T S_x \Phi + RK = 0$ into (6.27b) yields $S_{xc_{\{1,:\}}}^T = 0$. Subsequently, the second equality of (6.27a) is as follows:

$$S_{xc_{\{1,:\}}}^T \Phi = S_{xc_{\{2,:\}}}^T. \tag{6.27c}$$

Inserting $S_{xc_{\{1,:\}}}^T = 0$ into (6.27c) yields $S_{xc_{\{2,:\}}}^T = 0$, irrespective of $\Phi \in \mathbb{R}^{n_x \times n_x}$ being full rank or not. Similarly, examining the subsequent equality of (6.27a):

$$S_{xc_{\{i,:\}}}^T \Phi = S_{xc_{\{i+1,:\}}}^T, \; i \in \{2, \cdots, n_c - 1\}. \tag{6.27d}$$

By induction, substituting $S_{xc_{\{i,:\}}}^T = 0$ into (6.27d) yields $S_{xc_{\{i+1,:\}}}^T = 0$ for $i \in \{2, \cdots, n_c - 1\}$. Consequently, all entries of S_{xc}^T are zeros which implies $S_{xc}^T = 0$, that proves $B^T S_x \Phi + RK = 0$ is a sufficient condition.

Proof of necessity: To demonstrate $B^T S_x \Phi + RK = 0$ is a necessary condition for $S_{xc}^T = 0$, a contradiction argument is used. Suppose $B^T S_x \Phi + RK \neq 0$, based on the equality of (6.27b), $S_{xc_{\{1,:\}}}^T$ becomes as follows:

$$S_{xc_{\{1,:\}}}^T = B^T S_x \Phi + RK \neq 0. \tag{6.28}$$

If $S_{xc_{\{1,:\}}}^T \neq 0$, then $S_{xc}^T \neq 0$. Thus, this proves $B^T S_x \Phi + RK = 0$ is a necessary condition for $S_{xc}^T = 0$, which forms the second condition of Theorem 6.6.

Remark 6.7 Theorem 6.6 demonstrates that the extra control layer that satisfies the conditions (6.23) will not impact on the underlying robust output-feedback control law unless constraints are predicted to be active. Consequently, in normal operation, the properties of the original closed-loop dynamics are retained and the additional control layer solely handles the upcoming information.

Remark 6.8 The reader may wonder whether an alternative form of the cost function such as (6.20), where only the measures of $\underrightarrow{c}$ and $\underrightarrow{d}$ are considered, could void need of the conditions (6.23). Indeed, the perturbation $\underrightarrow{c}$ from (6.20) would become independent of x_k but, however, the preview term K_{cd} becomes sub-optimal. This is because the cost function (6.20) is ill-posed if the conditions are not satisfied, where there would be a mismatch between the predictions of the state and input with the actual closed-loop trajectories. Examples are provided in Sect. 6.6.2 in order to demonstrate the sub-optimal control law if the the term S_{cx} is simply neglected in the cost function (6.20).

6.5.2 Discussions on Stability and Feasibility

Nevertheless, when constraints are active, the robustness of the feedback closed-loop dynamics cannot be retained as the perturbation c_k then impacts the closed-loop. The nominal stability of the proposed MPC layer can be established based on the use of infinite horizons and invariant set. Assumes that the optimisation (6.8) is feasible at every sample k, then, there must exist a perturbation sequence $\underrightarrow{c}$ that ensures the trajectories of the closed-loop system always satisfy the constraints (6.13). Thus, recursive feasibility of the optimisation (6.8) is implicitly required for the stability guarantee of the MPC layer.

However, in presence of model uncertainties or unmodelled disturbance or ambitious constraint requirements, there might not exist a perturbation sequence $\underrightarrow{c}$ that satisfy the constraint set (6.15). In addition, the steady-state $\Delta u^s_{i|k}, x^s_{i|k}$ could become unreachable due to active constraints. Thus, such factors could cause the optimisation (6.8) infeasible. In the literature, there are many studies regarding the problems of infeasibility in MPC, for example, the use of constraint-softening strategies (Scokaert & Rawlings, 1999; Zeilinger et al., 2014), steady-state target calculator (Rawlings et al., 2008; Shead et al., 2010). Nonetheless, the focus of this chapter is on a simple MPC formulation on an output-feedback controller, thus, details regarding infeasibility issues of the MPC are not discussed further in this chapter. Instead, this chapter adopts a practical solution that whenever infeasibility occurs, the MPC layer backs off and the input is simply falling back to the pre-determined stabilising control law $\Delta\tilde{u}_k = K\tilde{x}_k$. Under the assumption that the original closed-loop is stable when the constraints are active, this method also ensures stability at the sacrifice of the performance that the available feed-forward information is not exploited.

6.5.3 Tuning of the MPC Layer

Since that the stabilising controller K is pre-determined, the weights Q, R that satisfy (6.23) can be computed by solving a linear matrix inequality (LMI) problem (Boyd, El Ghaoui, Feron, & Balakrishnan, 1994). The choice of preview horizon n_a can be chosen as large as possible given that the cost function (6.8) takes into account the full preview knowledge, whilst the control horizon n_c rests largely on preference of design. Nonetheless, a longer control horizon n_c can result in a larger maximal controlled admissible set S, thus implicitly lowering the chance of infeasibility. However, more computational power is required for a large n_c.

Remark 6.9 The reader may wonder whether the corresponding Q, R need not imply good performance. However, this is not the case as the inner loop is specially designed to meet frequency domain criteria in preference to time-domain criteria and hence implicitly the corresponding Q, R are appropriate.

Remark 6.10 The condition (6.23) only requires satisfaction of the weights Q, R within the cost function (6.8), regardless of the choice of preview and control horizon n_a, n_c.

6.6 Numerical Examples and Discussions

For the examples presented through this chapter, consider a SISO system given by:

$$y(s) = G(s)u(s) + G_{\mathrm{d}}(s)d(s) \tag{6.29a}$$

$$= \frac{0.2}{(s+0.25)(0.11s+1)}u(s) - \frac{0.5}{s+0.25}d(s), \tag{6.29b}$$

and a robust stabilising controller is given as follows:

$$u(s) = K(s)y(s) \tag{6.30}$$

$$= -\frac{11.40s + 4.08}{s}y(s). \tag{6.31}$$

Subsequently, with the sampling time $T = 0.2$s, the discrete-time closed-loop model is as follows:

$$\begin{bmatrix} x_{\mathrm{p}_{k+1}} \\ x_{K_{k+1}} \\ u_k \end{bmatrix} = \underbrace{\begin{bmatrix} 0.15 & -0.06 & 0 & -0.11 \\ 0.18 & 0.96 & 0 & -0.07 \\ -0.03 & 0.1614 & 1 & 0 \\ 0 & 0 & 0 & 1 \end{bmatrix}}_{A} \underbrace{\begin{bmatrix} x_{\mathrm{p}_k} \\ x_{K_k} \\ u_{k-1} \end{bmatrix}}_{x_k} + \underbrace{\begin{bmatrix} -0.11 \\ -0.07 \\ 0 \\ 1 \end{bmatrix}}_{B} \Delta u_k + \underbrace{\begin{bmatrix} -0.02 \\ 0.24 \\ 0 \\ 0 \end{bmatrix}}_{B_{\mathrm{d}}} d_k, \tag{6.32}$$

$$\Delta u_k = K x_k = \begin{bmatrix} -0.82 & 4.60 & 2.04 & -1 \end{bmatrix} x_k, \tag{6.33}$$

$$y_k = C x_k = \begin{bmatrix} 0.07 & -0.40 & 0 & 0 \end{bmatrix} x_k, \tag{6.34}$$

where the eigenvalues of the closed-loop model $A+BK$ are $0, 0.92$ and $0.47 \pm 0.27i$. Let the prefect preview knowledge available up to $n_a = 10$ samples and the control horizon is chosen as $n_c = 2$ samples.

6.6.1 Analysis of the Weights Selection

This section will investigate the importance of weights selection for formulating the MPC layer. Four weights, as listed in Table 6.1, are chosen for the cost function (6.8) based on the closed-loop model (6.32). The first and second sets of weights Q_1, R_1 and Q_2, R_2 both satisfy the condition (6.23). Thus, given that $S_{xc}^T = 0$, it is expected that the eigenvalues of the closed-loop system with the MPC layer remain unchanged, which are the same as those of the original closed-loop (6.32), as shown in Table 6.1. In contrast, the third set of weights Q_3, R_3 are selected arbitrarily and the eigenvalues of the modified closed-loop reveal that the performance and robustness properties are affected by the additional MPC layer, as expected. To further illustrate the performance of these weights selection, simulation examples with a step disturbance are shown in the following section.

6.6.2 Example 1: A Step Disturbance

As discussed in Sect. 6.3.4, the use of feed-forward control law (6.6) might result in a high demanding input effort to achieve good disturbance rejection. The perturbation

Table 6.1 Weights selection for the cost function (6.8) of the MPC layer

Case i	Q_i	R_i	Satisfying (6.23)	Eigenvalues
1	$\begin{bmatrix} 356.18 & -57.01 & -9.82 & 58.83 \\ -57.01 & 31.76 & 3.82 & -8.60 \\ -9.82 & 3.82 & 9.47 & -2.19 \\ 58.83 & -8.60 & -2.19 & 10 \end{bmatrix}$	3.34×10^{-15}	✓	0 0.92 $0.47 \pm 0.27i$
2	$\begin{bmatrix} 37.39 & 2.51 & 4.00 & 5.81 \\ 2.51 & 2.34 & 0.71 & 0.44 \\ 4.00 & 0.71 & 0.52 & 0.63 \\ 5.81 & 0.44 & 0.63 & 0.90 \end{bmatrix}$	2.40×10^{-15}	✓	0 0.92 $0.47 \pm 0.27i$
3	$\begin{bmatrix} 1 & 0 & 0 & 0 \\ 0 & 1 & 0 & 0 \\ 0 & 0 & 1 & 0 \\ 0 & 0 & 0 & 1 \end{bmatrix}$	1	✗	0.16 0.38 $0.90 \pm 0.06i$

The eigenvalues represents the poles of the closed-loop with the MPC layer $A + B(K - K_{cx})$

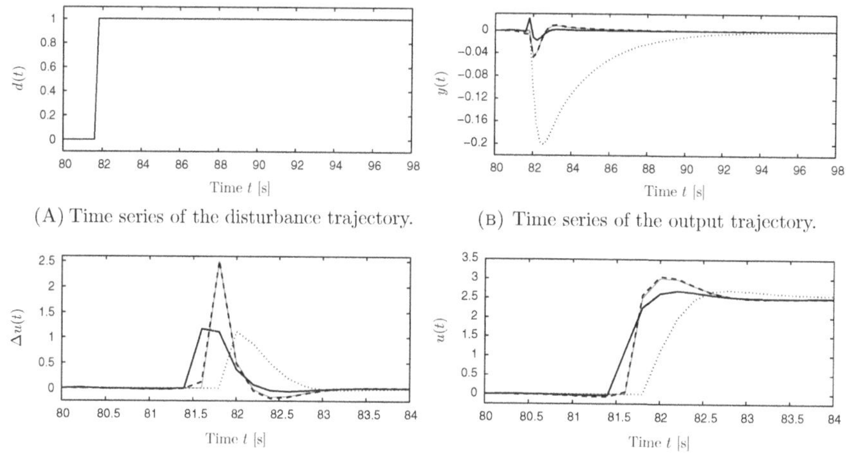

(A) Time series of the disturbance trajectory.

(B) Time series of the output trajectory.

(C) Time series of the input rate trajectory.

(D) Time series of the input trajectory.

Fig. 6.2 Time series of the disturbance, output, input rate and input trajectories. For **b, c** and **d**, dotted line denotes feedback controller (6.2), solid thin line represents feed-forward and feedback control law (6.6), dashed line is the MPC layer with Q_1, R_1 and solid thick line is the MPC layer with Q_2, R_2

$c_k = K_{cd}\underset{\rightarrow}{d}$ that exploits the preview knowledge can be tuned for improving the transition between the current input and the desired input trajectory. Consider a step disturbance which is perfectly available to the MPC layer, the input rate, input and output trajectories under a constraint-free case are shown in Fig. 6.2. The MPC layer with the weighs Q_1, R_1 performs similar to the feed-forward control law (6.6) in Fig. 6.2c, d and b. However, the MPC layer with Q_2, R_2 has a similar input rate effort as the original feedback controller in Fig. 6.2c but much better disturbance rejection as shown in Fig. 6.2b. This is because less penalties on Q_1 leads to a less aggressive input perturbation c_k. The results from this section demonstrate that the preview knowledge handling capability of the MPC layer can be tuned with preference of design.

The reader may wonder the arbitrary choice of Q_3, R_3 that does not satisfy the conditions (6.23) still gives a reasonable and stable closed-loop performance. However, the preview control law $c_k = -K_{cd}\underset{\rightarrow}{d}$ with the weights Q_3, R_3 might result in a sub-optimal behaviour, where the prediction of the closed-loop state is not consistent with actual closed-loop behaviour since that the cost function (6.8) does not embed any knowledge of the original controller (6.2). This sub-optimality also applies to the alternative cost function (6.20), where the S_{cx} terms is simply neglected and the weights are chosen arbitrarily. To illustrate the sub-optimality, consider the same setting as in Fig. 6.3 but with the MPC layer with Q_3, R_3, Fig. 6.3b demonstrates that there is mismatch between the prediction $\underset{\rightarrow}{c}$ and the actual behaviour of c_k. Additionally, the perturbation c_k does not converge to zero after sample 6, thus, the perturbation c_k affected the steady-state of the input. In contrast, Fig. 6.3a illustrates

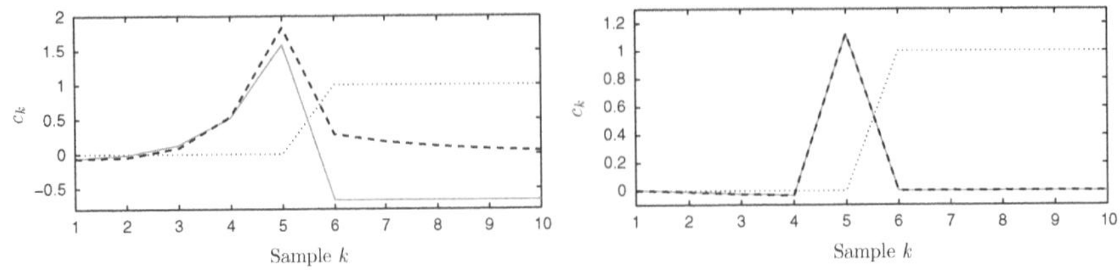

(A) The sub-optimal response of the MPC layer with Q_3, R_3.

(B) The optimal response of the MPC layer with Q_1, R_1.

Fig. 6.3 Prediction of the perturbation sequence $\underset{\to k}{c}$ at sample time 1 (dashed line) based on the step disturbance (dotted line) and the actual behaviour of the perturbation c_k (solid line)

the prediction of the perturbation sequence $\underset{\to k}{c}$ at sample time 1 is consistent with the actual behaviour of c_k. Thus, for the cost function (6.8) remains optimal, the corresponding weights need to satisfy condition (6.23).

6.6.3 *Example 2: A Time-Varying Disturbance and Constraints*

Consider a time-varying disturbance $d_k \in \{-16, 16\}$, the bounds on input, input rate and output can be computed by the steady-state target calculator (6.5) for formulating the maximal controlled admissible set (6.15), defined as follows:

$$\begin{bmatrix} x_k^s \\ \Delta u_k^s \end{bmatrix} = \left[-0.34 \; -0.06 \; 0 \; 2.50 \; 0\right]^T d_k, \tag{6.35a}$$

$$y_k^s = Cx_k^s = 0. \tag{6.35b}$$

Thus, only the input needs to be bounded as $u \in \{-40, 40\}$ as discussed in Assumption 6.4.2.

Consider the time-varying disturbance as shown in Fig. 6.4a, the response of the closed-loop system (6.32) with the feedback controller (6.30) is illustrated in Fig. 6.4b, c demonstrates the input rate operating near the input rate limits, assuming the input rate limit of the actuators as $\Delta u \in \{-8, 8\}$. Based on this closed-loop system model (6.32), a MPC layer can be constructed with the weights Q_2, R_2 in this example and the input rate is tuned moderately, as shown in Fig. 6.4c, which has similar input activity as the original feedback controller. By using the similar input effort, Fig. 6.4b illustrates the superior performance on disturbance rejection of the proposed MPC layer, assuming prefect advance knowledge and no modelling error.

To clearly demonstrate the constraint handling capability, the output is assumed to be bounded as $y \in \{-0.06, 0.06\}$. Figure 6.5 shows the performance of the proposed MPC structure with and without the output constraint requirement. Figure 6.5b, c

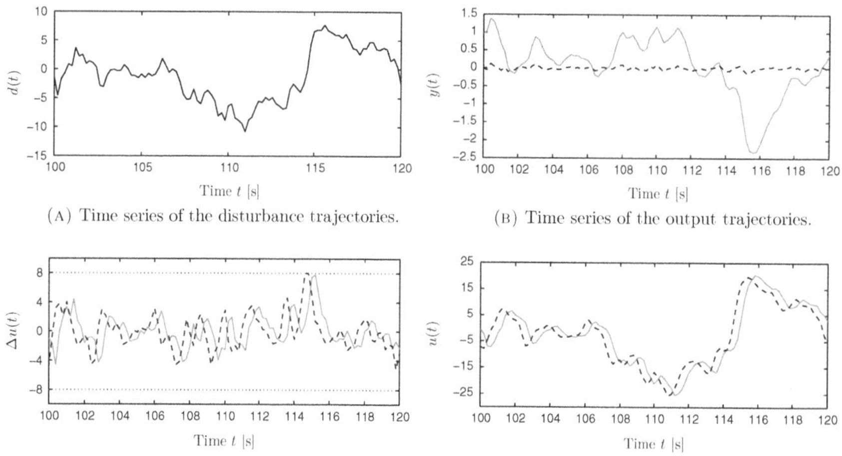

(A) Time series of the disturbance trajectories.

(B) Time series of the output trajectories.

(C) Time series of the input rate trajectories.

(D) Time series of the input trajectories.

Fig. 6.4 An example without the output constraint. Time series of the disturbance, output, input rate and input trajectories. For **b**, **c** and **d**, solid line denotes the feedback controller, whilst dashed line represents the proposed MPC layer structure and dotted line is the constraint

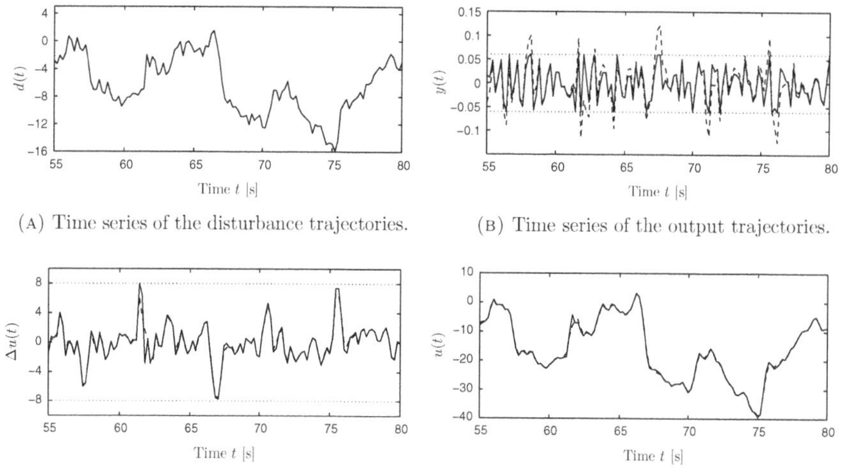

(A) Time series of the disturbance trajectories.

(B) Time series of the output trajectories.

(C) Time series of the input rate trajectories.

(D) Time series of the input trajectories.

Fig. 6.5 An example with the output constraint, Time series of the disturbance, output, input rate and input trajectories. For **b**, **c** and **d**, solid line denotes the feedback controller, whilst dashed line represents the proposed MPC layer structure and dotted line is the constraint reveal that the output-constrained MPC layer provides an effective control input for the closed-loop system with the knowledge of the bounds on the output.

In conclusion, the MPC layer brings immediate benefits to many applications with an existing feedback controller. For example, the perturbation from the MPC layer that handles the preview knowledge and constraints can be tuned systematically regarding the capability of the input actuators. Additionally, the separate nature of the MPC layer enables fair and transparent performance comparison between the original controller and the additional benefits from utilising the feed-forward information and constraint management.

6.7 Summary

This chapter has presented an MPC layer design where the feed-forward measurements and constraint handling capability are incorporated into a known output-feedback controller, together with the conditions to ensure the former does not interfere with the closed-loop dynamics provided by the latter. Two key benefits of the proposed MPC layer design were discussed, namely: (i) the robustness and frequency-domain properties of the nominal output-feedback controller are retained in the unconstrained case and such robust properties are likely to extend to the constrained case; (ii) the separate nature of the proposed structure offers a clear and transparent framework to quantify the benefits. Several simulation studies have been provided to illustrate the efficacy of the MPC layer design and also the importance of weight selection to satisfy the proposed conditions.

In the coming chapter, the MPC layer is formulated based upon the baseline feedback controller, which is discussed in Sect. 2.8. And the benefits of the uses of the upcoming wind information are examined.

References

Boyd, S., El Ghaoui, L., Feron, E., & Balakrishnan, V. (1994). *Linear matrix inequalities in system and control theory*. Society for Industrial and Applied Mathematics.

Carrasco, D. S., & Goodwin, G. C. (2011a). Feedforward model predictive control. *Annual Reviews in Control*, *35*(2), 199–206.

Carrasco, D. S., & Goodwin, G. C. (2011b). Preview and feedforward in model predictive control: A preliminary robustness analysis. In *IFAC World Congress* (Vol. 18, pp. 185–190).

Clarke, D., Mohtadi, C., & Tus, P. (1987). Generalized predictive control–Part I. The basic algorithm. *Automatica*, *23*(2), 137–148.

Dughman, S., & Rossiter, J. (2015). A survey of guaranteeing feasibility and stability in MPC during target changes. In *IFAC ADCHEM 2015*.

Gilbert, E., & Tan, K. (1991). Linear systems with state and control constraints: The theory and application of maximal output admissible sets. *IEEE Transactions on Automatic Control*, *36*(9), 1008–1020.

Goodwin, G., Carrasco, D., & Mayne, D. (2011). Preview and feedforward in model predictive control: Conceptual and design issues. In *IFAC World Congress* (pp. 5555–5560).

Limon, D., Alvarado, I., Alamo, T., & Camacho, E. (2008). MPC for tracking piecewise constant references for constrained linear systems. *Automatica*, *44*(9), 2382–2387.

Muske, K. R., & Rawlings, J. B. (1993). Model predictive control with linear models. *AIChE Journal*, *39*, 262–287.

Pannocchia, G., & Rawlings, J. B. (2003). Disturbance models for offset-free model predictive control. *AIChE Journal*, *49*(2), 426–437.

Pluymers, B., Rossiter, J., Suykens, J., & De Moor, B. (n.d.). The efficient computation of polyhedral invariant sets for linear systems with polytopic uncertainty. In *Proceedings of the 2005, American Control Conference* (pp. 804–809). IEEE.

Rawlings, J. B., Bonné, D., Jørgensen, J. B., Venkat, A. N., & Jørgensen, S. B. (2008). Unreachable setpoints in model predictive control. *IEEE Transactions on Automatic Control*, *53*(9), 2135–2141.

Rossiter, J. A. (2003). *Model-based predictive control: A practical approach*. Boca Raton: CRC Press.

Rossiter, J. A., & Grinnell, B. G. (1995). Extended input horizon generalized predictive control—A GPC algorithm with better tracking. In *3rd IEEE Symposium on New Directions in Control and Automation*. Cyprus.

Rossiter, J. A., & Grinnell, B. G. (1996). Improving the tracking of generalized predictive control controllers. In *Proceedings of the Institution of Mechanical Engineers*.

Rossiter, J. A., Kouvaritakis, B., & Rice, M. J. (1998). A numerically robust state-space approach to stable-predictive control strategies. *Automatica*, *34*(1), 65–73.

Rossiter, J. A., & Valencia-Palomo, G. (2009). Feed forward design in MPC. In *European Control Conference*.

Scokaert, P. O. M., & Rawlings, J. B. (1999). Feasibility issues in linear model predictive control. *AIChE Journal*, *45*(8), 1649–1659.

Shead, L., Muske, K., & Rossiter, J. (2008). Conditions for which MPC fails to converge to the correct target. *IFAC Proceedings Volumes*, *41*(2), 6968–6973.

Shead, L. R. E., Muske, K. R., & Rossiter, J. A. (2010). Conditions for which linear MPC converges to the correct target. *Journal of Process Control*, *20*(10), 1243–1251.

Valencia-Palomo, G., Rossiter, J. A., & López-Estrada, F. R. (2014). Improving the feedforward compensator in predictive control for setpoint tracking. *ISA Transactions*, *53*(3), 755–766.

Wang, L., & Rossiter, J. A. (2008). Disturbance rejection and set-point tracking of sinusoidal signals using generalized predictive control. In *Proceedings of CDC* (Vol. 1, pp. 4079–4084).

Zeilinger, M. N., Morari, M., & Jones, C. N. (2014). Soft constrained model predictive control with robust stability guarantees. *IEEE Transactions on Automatic Control*, *59*(5), 1190–1202.

Chapter 7
Feed-Forward Model Predictive Control Layer on Wind Turbines

7.1 Overview and Contribution

The use of upstream wind measurements has motivated the development of blade-pitch preview controllers to improve rotor speed tracking and structural load reduction beyond that achievable via conventional feedback control. Such preview controllers, typically based upon model predictive control (MPC) for its constraint handling properties, alter the closed-loop dynamics of the existing blade-pitch feedback control system. This can result in a deterioration of the robustness properties and performance of the existing feedback control system. Furthermore, performance gains from utilising the upcoming real-time measurements cannot be easily distinguished from the feedback control, making it difficult to formulate a clear business case for the use of preview control. Therefore, the aim of this work is to formulate a modular MPC layer on top of a given output feedback blade-pitch controller, with a view to retaining the closed-loop robustness and frequency-domain performance of the latter. We derive a key result that proves that the proposed modular MPC layer handles real-time advance measurements and impacts the existing closed-loop system if and only if constraints are violated. The separate nature of the proposed controller structure enables clear and transparent quantification of the benefits gained by using preview control, beyond that of the underlying feedback controller. This is illustrated by results obtained from high-fidelity closed-loop turbine simulations, showing the performance comparison between a nominal feedback controller and an additional MPC-based preview controller. The proposed control scheme incorporating knowledge of the oncoming wind and constraints achieved significant 43 and 30% reductions in the rotor speed and flap-wise blade moment standard deviations, respectively. Additionally, the chance of constraint violations on the rotor speed decreased remarkably from 2.15 to 0.01%, compared to the nominal controller.

The remainder of this chapter is structured as follows. Section 7.2 presents an introduction and motivation of this chapter. In Sect. 7.3, the modelling aspect of the blade pitch control problem, including model disturbance, and the detail of the nominal output feedback controller are discussed. This is followed in Sect. 7.4 by

W. H. (Alan) Lio, *Blade-Pitch Control for Wind Turbine Load Reductions*, Springer Theses, https://doi.org/10.1007/978-3-319-75532-8_7

a formulation of a predictive control layer. In Sect. 7.5, details of simulation environments and tuning of the MPC layer will be discussed. In Sect. 7.6, simulation results on a high-fidelity wind turbine under various wind cases are demonstrated, showing the benefits of deploying the proposed control layer on top of the output-feedback controller. Section 7.7 concludes this chapter with a summary and an overview of future work.

7.2 Introduction

In recent years, a growing body of research has emerged, seeking to utilise real-time measurement of the approaching wind field from sensing devices for feed-forward design to further improve the performance of blade pitch control systems. Early adoption of feed-forward control in wind turbines that focused on turbine speed regulation was reported by Kodama, Matsuzaka, Tuchiya, and Arinaga (1999), in which the feed-forward control strategy was based on the hub-height wind measurement taken 40 m in front of the rotor by an anemometer on a free standing tower. Light detection and ranging (LIDAR) devices, employed by Harris, Hand, and Wright (2005) and numerous subsequent authors (e.g. Bossanyi et al., 2012; Dunne, Pao, Wright, Jonkman, Kelley, & Simley 2011; Schlipf, Fischer, & Carcangiu, Schlipf et al. (2010)), demonstrated the impact on the performance of the CPC in regulating rotor speed and mitigating tower load by exploiting preview information of the approaching wind field. Lately, a number of authors (e.g. Dunne, Pao, Wright, Jonkman, and Kelley (2011); Laks, Pao, Wright, et al., 2011; Schlipf, Schuler, Grau, & Martin, 2010; Wang, Johnson, Wright, & Wright, 2012) investigated the use of upcoming wind measurements with IPC, aiming to attenuate unbalanced loads on the rotor and blade loads. Several authors (e.g. Laks, Pao, Simley, et al., 2011; Lio, Rossiter, & Jones, 2014; Spencer, Stol, Unsworth, Cater, & Norris, 2013) employed model predictive control (MPC), for its constraint handling features, to design preview CPC and IPC controllers. The results by these authors demonstrated the efficacy of preview MPC design for turbine speed regulation and flap-wise blade load reduction.

The majority of preview MPC studies in wind turbines used a standard MPC approach where their shortcomings are that the robustness and closed-loop frequency-domain properties are usually not well considered in the time-domain design. As the loads on turbine blades predominately exist at the harmonics of the blade rotational frequency, thus, it is more intuitive to design a robust closed-loop feedback controller in the frequency-domain to attenuate such loads. Consequently, the modular MPC layer design proposed in Chap. 6 is adopted in this chapter and formulated based on a known robust output-feedback blade-pitch controller where the MPC layer handles constraints and upcoming wind measurements whilst retaining the robustness properties of the existing closed-loop. The separate nature of this MPC layer is important from an industry perspective, since the feed-forward control can be implemented without replacing the existing feedback controller. Furthermore, it provides a clear framework to quantify the benefits of the use of advance real-time measurements and constraint handling capability over the nominal output-feedback strategy, which forms the essence of this chapter.

7.3 Wind Turbine Modeling and Nominal Robust Feedback Compensator

This section gives a brief background of wind turbine modelling including model disturbances and details of the chosen robust feedback controllers that are later employed by the proposed MPC layer formulation.

Remark 7.1 The wind turbine model employed in this chapter is largely based on Sect. 4.3 in Chap. 4. In this chapter, the difference is that the rotor dynamics is taken into account and the disturbance modelling is also included.

7.3.1 Wind Turbine Modelling

A typical wind turbine blade-pitch control system architecture for above-rated conditions is shown in Fig. 7.1. The CPC regulates the rotor speed $\omega(t)$ by adjusting the collective pitch angle signal, whilst the IPC attenuates loads by providing additional pitch signals to the collective pitch angle in response to flap-wise blade root bending moment signals. To isolate the action of the IPC from that of the CPC, it is convenient to define the pitch angles and blade moments as follows:

$$\begin{bmatrix} \theta_1^c(t) \\ \theta_2^c(t) \\ \theta_3^c(t) \end{bmatrix} := \begin{bmatrix} \bar{\theta}^c(t) + \tilde{\theta}_1^c(t) \\ \bar{\theta}^c(t) + \tilde{\theta}_2^c(t) \\ \bar{\theta}^c(t) + \tilde{\theta}_3^c(t) \end{bmatrix}, \quad \begin{bmatrix} M_1(t) \\ M_2(t) \\ M_3(t) \end{bmatrix} := \begin{bmatrix} \bar{M}(t) + \tilde{M}_1(t) \\ \bar{M}(t) + \tilde{M}_2(t) \\ \bar{M}(t) + \tilde{M}_3(t) \end{bmatrix}, \tag{7.1}$$

where $\tilde{\theta}_i^c(t)$, for $i = \{1, 2, 3\}$, represent the perturbations in blade pitch demand from collective pitch angle signal $\bar{\theta}^c(t)$. Similarly, $\tilde{M}_i(t)$, for $i = \{1, 2, 3\}$, are the perturbations in flap-wise blade root bending moments, obtained by filtering out the average moment $\bar{M}(t)$ from the measurements $M_{1,2,3}(t)$. This structure is commonly used to separate the action of the IPC from that of the CPC (e.g. Bossanyi, 2003; Geyler & Caselitz, 2008; Lu, Bowyer, & Jones, 2015; Selvam, Kanev, van Wingerden, van Engelen, & Verhaegen, 2009). The relationship between collective pitch input $\bar{\theta}^c$ and rotor speed output ω can be modelled by a transfer function $G_{\omega\theta} \in \mathcal{R}$ obtained by linearising the turbine dynamics around the operating conditions. In a similar fashion, the transfer function $G_{M\theta} \in \mathcal{R}$ relating each flap-wise blade bending moment output $\tilde{M}_i$ to additional pitch inputs $\tilde{\theta}_i^c$ for $i = \{1, 2, 3\}$ can also be found. For simplicity, it is assumed that there is no coupling between the CPC and IPC loops from the tower dynamics. These transfer functions are defined as follows:

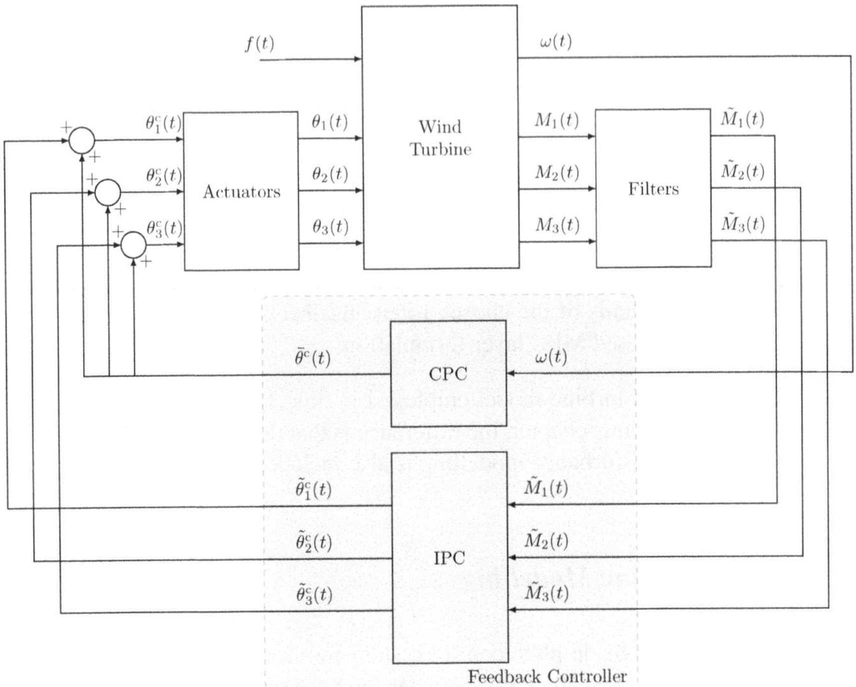

Fig. 7.1 System architecture of a wind turbine blade-pitch control system, combining collective pitch control (CPC) and individual pitch control (IPC). The CPC regulates rotor speed while the IPC attenuates perturbations in the flap-wise root bending moments on each blade. Additional inputs to the turbine, such as wind loading and generator torque, are accounted for in the term $f(t)$

$$G_{\omega\theta}(s) := G_{\mathrm{a}}(s)G_{\mathrm{r}}(s), \tag{7.2a}$$

$$G_{M\theta}(s) := G_{\mathrm{a}}(s)G_{\mathrm{b}}(s)G_{\mathrm{bp}}(s), \tag{7.2b}$$

where G_{r}, G_{b}, $G_{\mathrm{a}} \in \mathcal{R}$ describe the dynamics of rotor, blade and actuator, respectively, whilst $G_{\mathrm{bp}} \in \mathcal{R}$ is a band-pass filter that is included in order to remove the low and high frequency contents of the blade root bending moment measurement signals, obtained from strain-gauge sensors. These transfer functions are defined as follows:

$$G_{\mathrm{r}}(s) := \frac{\partial \omega}{\partial \theta} \frac{1}{\tau_{\mathrm{r}} s + 1}, \tag{7.3a}$$

$$G_{\mathrm{b}}(s) := \frac{\partial M_{\mathrm{flap}}}{\partial \theta} \frac{(2\pi f_{\mathrm{b}})^2}{s^2 + 4\pi f_{\mathrm{b}} D_{\mathrm{b}} s + (2\pi f_{\mathrm{b}})^2}, \tag{7.3b}$$

$$G_{\mathrm{a}}(s) := \frac{1}{\tau_{\mathrm{a}} s + 1}, \tag{7.3c}$$

Table 7.1 Model parameter of $G(s)$ and $G_d(s)$ (7.6)

Parameters	Values	Units	Parameters	Values	Units
$\frac{\partial \omega}{\partial \theta}$	−0.84	rpm deg^{-1}	$\frac{\partial M_{\text{flap}}}{\partial \theta}$	-1.50×10^6	Nm deg^{-1}
τ_{r}	4.00	s	f_{b}	0.70	Hz
D_{b}	0.47	–	τ_{a}	0.11	s
f_{h}	0.80	Hz	f_{l}	0.014	Hz

$$G_{\text{bp}}(s) := \frac{2\pi f_{\text{h}}}{s^2 + 2\pi(f_{\text{h}} + f_{\text{l}})s + 4\pi^2 f_{\text{h}} f_{\text{l}}}, \tag{7.3d}$$

where $\frac{\partial \omega}{\partial \theta}$, $\tau_{\text{r}} \in \mathbb{R}$ denote the variation of aerodynamic torque to pitch angle and the time constant of the rotor dynamics, respectively, whilst $\frac{\partial M_{\text{flap}}}{\partial \theta}$, D_{b}, $f_{\text{b}} \in \mathbb{R}$ represent the variation of flap-wise blade root bending moment to pitch angle, blade damping ratio and natural frequency of first blade mode, respectively. $\tau_{\text{a}} \in \mathbb{R}$ denotes the time constant of the pitch actuator whilst f_{h}, $f_{\text{l}} \in \mathbb{R}$ represent the upper and lower cut-off frequencies of the band-pass filter, respectively. The values of those parameters are listed in Table 7.1. The dynamics of rotor speed (7.3a) and pitch actuator (7.3c) are approximated as first-order systems respectively whilst the blade dynamics (7.3b) and band-pass filter (7.3d) are modelled as second-order systems. Note that the high-fidelity wind turbine employed for simulation purposes in this study operates across above-rated wind conditions and the parameters $\frac{\partial \omega}{\partial \theta}$ and $\frac{\partial M_{\text{flap}}}{\partial \theta}$ vary based on operating wind conditions. For the linear models (7.3), a fixed set of parameters were obtained from linearisation of the simulation turbine model operating at 18 ms^{-1}, chosen since this value is close to the centre of the range of wind speeds covering above-rated wind conditions.

7.3.2 Disturbance Modelling

The rotor and blade are subjected to a temporally varying and spatially distributed wind field and in many studies, the feed-forward control assumes only a few points of wind measurement across the rotor disk to estimate the effective wind speed at the rotor and blade. Given the fact that the blade and rotor loads vary along the span of the blades, owing to the wind conditions and blade geometry, more wind measurements across the entire rotor plane will inevitably provide improved estimation of such loads. A number of studies demonstrated the feasibility of estimating the wind-field from a few point measurements taken upstream of the turbine (e.g. Raach, Schlipf, Haizmann, & Cheng, 2014; Towers & Jones, 2016). Since this is a non-trivial problem, the issue of wind-field estimation is not considered in this thesis. Instead, this chapter assumes the approaching stream-wise wind speeds are known apriori perfectly, and the focus of this chapter is to design a control algorithm that utilises such preview information.

The disturbance trajectories of rotor speed ω_d, and flap-wise blade bending moment $\tilde{M}_{d_i}$, for $i \in \{1, 2, 3\}$, caused by the approaching wind at sample time k, are defined as follows:

$$\omega_d(k) := \sum_{l,\phi} \frac{\partial \omega}{\partial v}(\bar{v}, l) v(l, \phi, k), \tag{7.4a}$$

$$\tilde{M}_{d_i}(k) := \sum_{l,\phi} \frac{\partial M_{\text{flap}}}{\partial v}(\bar{v}, l) v(l, \phi, k), \quad i = 1, 2, 3, \tag{7.4b}$$

where $v(l, \phi, k) \in \mathbb{R}$ denote the stream-wise wind speed measurements where $l, \phi \in \mathbb{R}$ represent the radial and angular co-ordinates across the rotor disk whilst $\bar{v} \in \mathbb{R}$ denote the averaged wind speed of the measurements. The variations in rotor speed and blade bending moment with respect to the wind are denoted as $\frac{\partial \omega_d}{\partial v}, \frac{\partial M_d}{\partial v} \in \mathbb{R}$. The rotor speed response ω to wind-induced disturbance ω_d is modelled as a first-order transfer function $G_{\omega\omega_d} \in \mathcal{R}$, whilst the response of flap-wise blade root bending moment $\tilde{M}_i$ to wind-induced disturbance $\tilde{M}_{d_i}$, for $i \in \{1, 2, 3\}$, is modelled as $G_{MM_d} \in \mathcal{R}$:

$$G_{\omega\omega_d}(s) := \frac{1}{\tau_r s + 1}, \tag{7.5a}$$

$$G_{MM_d}(s) := \frac{(2\pi f_b)^2}{s^2 + 4\pi f_b D_b s + (2\pi f_b)^2} G_{bp}(s), \tag{7.5b}$$

where the parameters are listed in Table 7.1. Combining (7.2) and (7.5), the overall transfer function models $G \in \mathcal{R}^{4\times4}$ and $G_d \in \mathcal{R}^{4\times4}$ can be represented as follows:

$$\begin{aligned}
\begin{bmatrix} \omega(s) \\ \tilde{M}_1(s) \\ \tilde{M}_2(s) \\ \tilde{M}_3(s) \end{bmatrix} &= \underbrace{\begin{bmatrix} G_{\omega\theta}(s) & 0 & 0 & 0 \\ 0 & G_{M\theta}(s) & 0 & 0 \\ 0 & 0 & G_{M\theta}(s) & 0 \\ 0 & 0 & 0 & G_{M\theta}(s) \end{bmatrix}}_{G(s)} \begin{bmatrix} \bar{\theta}^c(s) \\ \tilde{\theta}_1^c(s) \\ \tilde{\theta}_2^c(s) \\ \tilde{\theta}_3^c(s) \end{bmatrix} \\
&+ \underbrace{\begin{bmatrix} G_{\omega\omega_d}(s) & 0 & 0 & 0 \\ 0 & G_{MM_d}(s) & 0 & 0 \\ 0 & 0 & G_{MM_d}(s) & 0 \\ 0 & 0 & 0 & G_{MM_d}(s) \end{bmatrix}}_{G_d(s)} \begin{bmatrix} \omega_d(s) \\ \tilde{M}_{d_1}(s) \\ \tilde{M}_{d_2}(s) \\ \tilde{M}_{d_3}(s) \end{bmatrix}.
\end{aligned} \tag{7.6}$$

Equivalently, the model can be described in a discrete-time state-space form since the discrete-time model is more convenient in the MPC framework:

$$\begin{aligned} x_{\mathrm{p}}(k+1) &= A_{\mathrm{p}} x_{\mathrm{p}}(k) + B_{\mathrm{p}} u(k) + B_{\mathrm{d_p}} d(k), \\ y(k) &= C_{\mathrm{p}} x_{\mathrm{p}}(k), \end{aligned} \tag{7.7a}$$

where

$$u(k) = [\bar{\theta}^{\mathrm{c}}(k), \tilde{\theta}_1^{\mathrm{c}}(k), \tilde{\theta}_2^{\mathrm{c}}(k), \tilde{\theta}_3^{\mathrm{c}}(k)]^T, \tag{7.7b}$$

$$y(k) = [\omega(k), \tilde{M}_1(k), \tilde{M}_2(k), \tilde{M}_3(k)]^T, \tag{7.7c}$$

$$d(k) = [\omega_d(k), \tilde{M}_{d_1}(k), \tilde{M}_{d_2}(k), \tilde{M}_{d_3}(k)]^T, \tag{7.7d}$$

and the state vector $x_{\mathrm{p}} \in \mathbb{R}^{n_{x_{\mathrm{p}}}}$ of the model is a collection of variables that characterises the dynamics of the transfer functions G and G_d that map the input vector $u \in \mathbb{R}^{n_u}$ and disturbance vector $d \in \mathbb{R}^{n_d}$ into the output vector $y \in \mathbb{R}^{n_y}$. The subscript p denotes the plant.

7.3.3 Nominal Robust Feedback Controller

The focus of this chapter is to design the MPC layer algorithm on top of a nominal output-feedback controller. The chosen feedback controller $K \in \mathcal{R}^{4\times4}$, consisting of CPC $K_{\theta\omega} \in \mathcal{R}$ and IPC $K_{\theta M} \in \mathcal{R}$ is defined as follows:

$$\begin{bmatrix} \bar{\theta}^{\mathrm{c}}(s) \\ \tilde{\theta}_1^{\mathrm{c}}(s) \\ \tilde{\theta}_2^{\mathrm{c}}(s) \\ \tilde{\theta}_3^{\mathrm{c}}(s) \end{bmatrix} = \underbrace{\begin{bmatrix} K_{\theta\omega}(s) & 0 & 0 & 0 \\ 0 & K_{\theta M}(s) & 0 & 0 \\ 0 & 0 & K_{\theta M}(s) & 0 \\ 0 & 0 & 0 & K_{\theta M}(s) \end{bmatrix}}_{K(s)} \begin{bmatrix} \omega(s) \\ \tilde{M}_1(s) \\ \tilde{M}_2(s) \\ \tilde{M}_3(s) \end{bmatrix}, \tag{7.8}$$

where $K_{\theta\omega}, K_{\theta M} \in \mathcal{R}$ are chosen to be the baseline CPC and IPC controllers in Sect. 2.8. Notice that these feedback controllers yield a robust stability margin $b(GK, I) = 0.53$.

With respect to $K_{\theta M}$, a variety of IPC strategies exists in the literature, for example, Coleman transform-based control (Bossanyi, 2003; Lu et al., 2015), Clarke transform-based control (Zhang, Cheng, & Chen, 2013) and single-blade control (Leithead, Neilson, & Dominguez, Leithead et al. 2009). The transform-based IPC techniques involve coordinate mappings on the pitch inputs which complicate the constraint formulation in MPC, where the linear constraint inequalities needed to be updated on-line at every sample, based on the prediction of azimuth angle. As discussed in Chap. 4, the performance differences between the various types of IPCs is negligible. Consequently, single-blade control IPC is employed in this chapter, where each blade is equipped with its own controller ($K_{\theta M}$) in response to a local blade load measurement. The diagonal structure of the controller (7.8) mirrors that of the plant model (7.6). Implicit in this structure is an assumption of no dynamic coupling between the fixed and rotating turbine structures. The simulation results in Lio, Jones, Lu, and Rossiter (2015) showed that a controller of the form (7.8) could be designed to be insensitive to such coupling by shaping the open-loop frequency response to have low gain at the tower frequency. Similar to the plant model, the feedback controller (7.8) has a discrete-time state-space realisation:

$$\begin{aligned} x_\kappa(k+1) &= A_\kappa x_\kappa(k) - B_\kappa y(k), \\ u(k) &= C_\kappa x_\kappa(k) - D_\kappa y(k), \end{aligned} \tag{7.9}$$

where the state vector $x_\kappa \in \mathbb{R}^{n_{x_\kappa}}$ is a collection of variables that characterises the dynamics of the controller K. The subscript κ denotes controller.

7.4 Formulating the MPC Layer to Wind Turbines

This section presents the design of the MPC layer to compliment the output-feedback blade-pitch controller (7.8) derived in the previous section. Notice that the MPC layer formulation was mostly discussed in Chap. 6. This section discusses the issues which arise from formulating the MPC layer on the wind turbine.

7.4.1 State-Space Representation of the Closed-Loop System Model

The closed-loop dynamic system model employed in the proposed MPC algorithm can be described by combining the linear wind turbine model (7.7) and controller (7.9), and the model is defined as follows:

$$\begin{bmatrix} x_{\mathrm{p}k+1} \\ x_{\kappa k+1} \\ u_k \end{bmatrix} = \underbrace{\begin{bmatrix} A_\mathrm{p} & 0 & B_\mathrm{p} \\ -B_\kappa C_\mathrm{p} & A_\kappa & 0 \\ 0 & 0 & I \end{bmatrix}}_{A} \underbrace{\begin{bmatrix} x_{\mathrm{p}k} \\ x_{\kappa k} \\ u_{k-1} \end{bmatrix}}_{x_k} + \underbrace{\begin{bmatrix} B_\mathrm{p} \\ 0 \\ I \end{bmatrix}}_{B} \Delta u_k + \underbrace{\begin{bmatrix} B_{\mathrm{d_p}} \\ 0 \end{bmatrix}}_{B_\mathrm{d}} d_k, \tag{7.10a}$$

$$\Delta u_k = \begin{bmatrix} -D_\kappa C_\mathrm{p} & C_\kappa & -I \end{bmatrix} x_k = K x_k, \tag{7.10b}$$

$$y_k = \begin{bmatrix} C_\mathrm{p} & 0 \end{bmatrix} x_k = C x_k. \tag{7.10c}$$

The states of the turbine model $x_{\mathrm{p}_k} \in \mathbb{R}^{n_{x_\mathrm{p}}}$ cannot be measured directly and the separate nature of the feedback controller prohibits direct access to the states of the controller $x_{\kappa k} \in \mathbb{R}^{n_{x_\kappa}}$, thus, observers employed to estimate these states are described by the following expressions:

$$\hat{x}_{\mathrm{p}_{k+1|k}} = A_{\mathrm{p}}\hat{x}_{\mathrm{p}_{k|k-1}} + B_{\mathrm{p}}u_k + B_{\mathrm{d_p}}d_k + L_{\mathrm{p}}(y_k - \hat{y}_{k|k-1}), \quad \hat{y}_{k|k-1} = C_{\mathrm{p}}\hat{x}_{\mathrm{p}_{k|k-1}}, \tag{7.11a}$$

$$\hat{x}_{K_{k+1|k}} = A_K\hat{x}_{K_{k|k-1}} - B_K y_k + L_K(u_k - \hat{u}_{k|k-1}), \quad \hat{u}_{k|k-1} = C_K\hat{x}_{K_{k|k-1}} - D_K y_k, \tag{7.11b}$$

where $\hat{x}_{\mathrm{p}} \in \mathbb{R}^{n_{x\mathrm{p}}}$ and $\hat{x}_K \in \mathbb{R}^{n_{xK}}$ denote the estimates of the state of wind turbine model and controller, respectively, and $L_{\mathrm{p}} \in \mathbb{R}^{n_{x\mathrm{p}} \times n_y}$ and $L_K \in \mathbb{R}^{n_{xK} \times n_u}$ are the observer gains. It is noted that due to the mismatch between the wind turbine linear model and the high-fidelity turbine model, a difference exists between $y \in \mathbb{R}^{n_y}$ and $\hat{y} \in \mathbb{R}^{n_y}$, whilst in contrast, $\hat{u} \in \mathbb{R}^{n_u}$ converges to $u \in \mathbb{R}^{n_u}$ only if there is no noise on the inputs.

7.4.2 MPC Layer Formulation

With the closed-loop state-space model (7.10), the MPC layer can be formulated based on the results from Sect. 6.4 in Chap. 6, where the MPC layer algorithm is as follows:

Algorithm 7.4.1 *(MPC layer)* At each sampling instant k:

1. Perform the optimisation

$$\min_{\underset{\to k}{c}} \underset{\to k}{c}^T S_c \underset{\to k}{c} + 2\underset{\to k}{c}^T (S_{cd}\underset{\to k}{d} + S_{xc}^T x_{0|k}), \tag{7.12a}$$

$$\text{s.t.} \quad \mathcal{M}x_{0|k} + \mathcal{N}\underset{\to k}{c} + \mathcal{V}\underset{\to k}{d} \leq b. \tag{7.12b}$$

2. Apply the first block element $c_k = c_{0|k}$ of the perturbation sequence $\underset{\to k}{c}$ within the embedded control law $\Delta\tilde{u}_k = K\tilde{x}_k + c_k$, where the perturbation sequence $\underset{\to k}{c}$ is the minimiser of the optimisation problem (7.12)

As discussed in Sect. 6.5.2, issues regarding feasibility with this form of the MPC layer (7.12) were not particularly considered. In case of infeasibility, the approach in Chap. 6 is to return to the feedback control law, $\Delta\tilde{u}_k = K\tilde{x}_k$, whilst the perturbation c_k is kept at zero. To overcome the infeasibility issues, soft-constraint formulation is introduced in the following section.

7.4.3 Soft Constraint Formulation

The constraints considered in this chapter are soft constraints to overcome the feasibility issues. Thus, constraints on the angles and rates of the blade-pitch actuators and the rotor speed with slack variables are defined as follows:

$$\theta_{\min} - \varepsilon^{\theta}_{i|k} \leq \theta_{i|k} \leq \theta_{\max} + \varepsilon^{\theta}_{i|k}, \qquad \forall i \geq 0, \tag{7.13a}$$

$$\dot{\theta}_{\min} - \varepsilon^{\dot{\theta}}_{i|k} \leq \dot{\theta}_{i|k} \leq \dot{\theta}_{\max} + \varepsilon^{\dot{\theta}}_{i|k}, \qquad \forall i \geq 0, \tag{7.13b}$$

$$\omega_{i|k} \leq \omega_{\max} + \varepsilon^{\omega}_{i|k}, \qquad \forall i \geq 0, \tag{7.13c}$$

where $\theta = [\theta_1, \theta_2, \theta_3]^T \in \mathbb{R}^3$, whereas $\theta_{\min}, \theta_{\max}, \dot{\theta}_{\min}, \dot{\theta}_{\max} \in \mathbb{R}^3$ denote the minimum and maximum of the angle and rate of the pitch actuators, respectively, whilst $\omega_{\max} \in \mathbb{R}$ represents the maximum rotor speed. Since the constraints on pitch actuators and rotor speed are state-constraints, thus, the slack variables $\varepsilon_{i|k} = [\varepsilon^{\theta}_{i|k}, \varepsilon^{\dot{\theta}}_{i|k}, \varepsilon^{\omega}_{i|k}]^T \in \mathbb{R}^{n_\varepsilon}$ are employed to soften the constraints to ensure the feasibility of the optimisation if necessary. To minimise the predictions of the slack variables $\underset{\rightarrow k}{\varepsilon} = [\varepsilon_{0|k}, ..., \varepsilon_{n_c-1|k}]^T \in \mathbb{R}^{n_\varepsilon n_c}$, a quadratic penalty together with l_1-norm penalty is added into the cost (7.12a) in Algorithm 7.4.1, as follows:

$$J = \underset{\rightarrow k}{c}^T S_c \underset{\rightarrow k}{c} + 2\underset{\rightarrow k}{c}^T S_{xc}^T x_{0|k} + 2\underset{\rightarrow k}{c}^T S_{cd} \underset{\rightarrow k}{d} + \underset{\rightarrow k}{\varepsilon}^T S_\varepsilon \underset{\rightarrow k}{\varepsilon} + L_\varepsilon^T \underset{\rightarrow k}{\varepsilon}. \tag{7.14}$$

The weights of the quadratic penalty $S_\varepsilon \in \mathbb{R}^{n_\varepsilon n_c \times n_\varepsilon n_c}$, that is a diagonal matrix, penalises the peak of constraint violations, whilst the weights of the l_1-norm penalty $L_\varepsilon \in \mathbb{R}^{n_\varepsilon n_c}$ penalises the total sum of violations (Scokaert & Rawlings, 1999). Details of tunings are provided in Sect. 7.5.3.

Subsequently, the inequalities (7.13) can be written in terms of the autonomous form (6.9), with $z_{i|k} = \Psi^i z_{0|k}$, as follows:

$$H\Psi^i z_{0|k} \leq h + h^{\varepsilon}_{i|k}, \quad \forall i \geq 0, \tag{7.15}$$

where the matrices are chosen as $Hz_{i|k} = [\theta_{i|k}, -\theta_{i|k}, \dot{\theta}_{i|k}, -\dot{\theta}_{i|k}, \omega_{i|k}]^T$,$h = [\theta_{\max}, -\theta_{\min}, \dot{\theta}_{\max}, -\dot{\theta}_{\min}, \omega_{\max}]^T$ and $h^{\varepsilon}_{i|k} = [\varepsilon^{\theta}_{i|k}, -\varepsilon^{\theta}_{i|k}, \varepsilon^{\dot{\theta}}_{i|k}, -\varepsilon^{\dot{\theta}}_{i|k}, \varepsilon^{\omega}_{i|k}]^T$. Given that the slack variables are not bounded for ensuring feasibility, a maximal controlled admissible set like (6.15) cannot be formed in the soft-constraint formulation. Thus, for a practical approach, this study formulates the inequalities by checking the constraints over twice the control horizon. The inequalities can be described by a set of suitable matrices ($\mathcal{M}, \mathcal{N}, \mathcal{V}, \mathcal{T}$ and b) as follows:

$$\mathcal{M}x_{0|k} + \mathcal{N}\underset{\rightarrow k}{c} + \mathcal{V}\underset{\rightarrow k}{d} - \mathcal{T}\underset{\rightarrow k}{\varepsilon} \leq b, \tag{7.16}$$

The matrices $\mathcal{M}, \mathcal{N}, \mathcal{V}, \mathcal{T}$ and b can be computed off-line and hence only the variables $x_{0|k}$, $\underset{\rightarrow k}{c}$, $\underset{\rightarrow k}{d}$ and $\underset{\rightarrow k}{\varepsilon}$ need to be updated on-line.

To sum up Sect. 7.4, the proposed MPC layer, at each sample k, employs the states $x_k = x_{0|k}$ of the closed-loop system and subsequently determines the optimal perturbation sequence $\underset{\rightarrow k}{c}$ that takes into account both upcoming measurements and constraints, by solving a constrained minimisation of the predicted cost (7.14) subject to soft constraints (7.16). This is summarised in Algorithm 7.4.2.

Algorithm 7.4.2 *(MPC layer with soft constraints)* At each sampling instant perform the constrained optimisation below. The first block element $c_k = c_{0|k}$ of the perturbation sequence is applied within the embedded control law, where $u_k = Kx_k + c_k$:

$$\min_{\underset{\to k}{c},\underset{\to k}{\varepsilon}} \underset{\to k}{c}^T S_c \underset{\to k}{c} + 2\underset{\to k}{c}^T (S_{cd} \underset{\to k}{d} + S_{xc}^T x_{0|k}) + \underset{\to k}{\varepsilon}^T S_\varepsilon \underset{\to k}{\varepsilon} + L_\varepsilon^T \underset{\to k}{\varepsilon}, \tag{7.17a}$$

$$\text{s.t.} \quad \mathcal{M}x_{0|k} + \mathcal{N}\underset{\to k}{c} + \mathcal{V}\underset{\to k}{d} - \mathcal{T}\underset{\to k}{\varepsilon} \leq b. \tag{7.17b}$$

7.4.4 Soft-Constrained MPC Layer and Stability

As discussed in Sect. 6.5 in Chap. 6, recursive feasibility is required for guarantee of stability of the MPC layer (7.12). The use of soft constraint solves the infeasibility issues in (7.12). However, the set of constraints (7.16) now relies upon the slack variables and hence, is not a maximal controlled admissible set. Thus, recursive feasibility of (7.17a) does not imply guarantee of stability.

The benefit of the use of soft constraints is that the slack variables become zero such that the optimisation problem (7.17a) becomes (7.12). If the slack variables are not zero, that implies the constraints (7.16) being relaxed. Given that the nature of the wind turbine problem, where the stability of the plant does not rely upon the constraint satisfaction, the relaxed constraints do not necessarily make the system become unstable. In addition, the constraint violations caused by the wind typically are temporary and lasting over a short period of time. Thus, as the cost (7.17a) penalises the slack variables heavily, the slack variables converge quickly whenever it is possible and then, the optimisation problem is back to the original form (7.12) with the slack variables being zero. In case of a huge wind disturbance, most modern wind turbine have a separate protocol to ensure the turbines stop operating, as discussed in Sect. 2.2 in Chap. 2.

7.5 Simulation Environment and Controller Tunings

In this section, the details of the turbine simulation are presented and also the estimation methods of the upcoming disturbance trajectories of the rotor speed and blade moments are discussed. Furthermore, it is followed by selections of the control horizon, the preview horizon and the constraints for the MPC layer.

7.5.1 Simulation Environment

The turbine model employed in this study is the NREL 5MW baseline turbine (Jonkman, Butterfield, Musial, & Scott, 2009) based on the FAST code (Jonkman

& Buhl Jr, 2005). This model is of much greater complexity than the model (7.6) employed for control design and includes flap-wise and edge-wise blade modes, in addition to the tower and drive train dynamics. In addition, the generator torque controller is assumed to be fixed in the above-rated conditions (Jonkman et al., Jonkman et al. 2009). Closed-loop simulations were performed under a set of representative and turbulent wind fields generated by the TurbSim code (Jonkman, 2009). These full-field three-dimensional wind data were characterised by mean wind speeds, turbulence settings and wind shear exponent. The TurbSim code simulated a time series of wind data at points in a two-dimensional 17-by-17 grid such that the sequence of grids march towards the rotor at a constant speed specified by the mean wind speed and under the assumption of Taylor's frozen turbulence hypothesis.

7.5.2 Future Measurements of Rotor Speed and Blade Disturbance

As discussed in Sect. 7.3.2, the issue of wind-field estimation is a non-trivial problem, this chapter assumes the disturbance trajectories of rotor speed and flap-wise blade bending moments are estimated based on the prefect stream-wise wind speed measurements in front of the turbine. To examine the accuracy of the estimated disturbance trajectories generated from (7.4), comparisons were made against the actual trajectories obtained from the non-linear turbine simulation, shown in Fig. 7.2. The time series of the disturbance trajectories of rotor speed deviation $\Delta\omega$ and flap-wise bending moment of blade 1 $\tilde{M}_1$ are illustrated in Fig. 7.2a, b and it reveals that the disturbance trajectories obtained from the linear model and non-linear turbine are almost identical. Figure 7.2c, d reveal the frequency spectra of the time series of both trajectories, which confirm that both trajectories are alike, as demonstrated by the similarity in the magnitudes at the dominant frequencies that is below 0.1 Hz for rotor speed and at 0.2 Hz for blade moment.

7.5.3 Choice of the MPC Horizons and Constraints

Effective constraint and feed-forward information handling relies on sensible choices of preview horizon n_a and control horizon n_c. The predictive controller should anticipate the upcoming disturbance far ahead enough to allow beneficial feed-forward compensation. The operating frequency of the MPC controller was 5 Hz which provided a satisfactory compromise between performance and computational burden. A preview horizon of $n_a = 15$ samples was found a reasonable choice in the present simulation setting; hence the preview horizon period was of duration three seconds. The choice of the control horizon n_c is dependent on the computational requirement

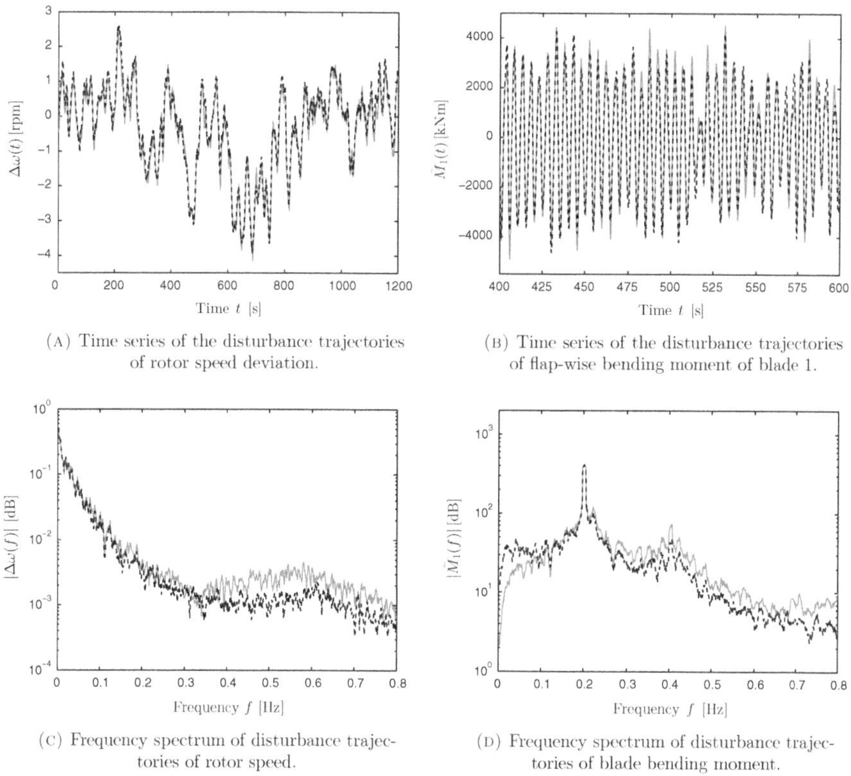

(A) Time series of the disturbance trajectories of rotor speed deviation.

(B) Time series of the disturbance trajectories of flap-wise bending moment of blade 1.

(C) Frequency spectrum of disturbance trajectories of rotor speed.

(D) Frequency spectrum of disturbance trajectories of blade bending moment.

Fig. 7.2 Comparison between the disturbance trajectories obtained from the linear model (dash line) and non-linear turbine (solid line). Simulation data was obtained under a turbulent wind field characterised by the mean speed of 18 ms^{-1} and turbulence intensity of 14%. Similar results were observed for the remaining blades

and related to feasibility issues of the MPC algorithm as discussed in Sect. 6.5.3. Thus, in this work, the control horizon n_c is also chosen as 15 samples.

The constraints considered in this study are the angles and rates of the blade pitch actuators and the rotor speed, as listed in Table 7.2. The pitch angle for each blade is bounded between 0° and 90°, whilst the limits on pitch rate for each blade are ±8 degrees per second. In addition, a constraint is also placed on the maximum rotor speed to avoid excessive loads on the generator. The value for the maximum rotor speed deviation is selected as 0.726 rpm (6%) from the rated speed,

Tuning of the weights of the MPC layer Q and R in the cost (7.14) is intuitive, as discussed in Sect. 6.5.3. In addition, the weights also need to satisfy the conditions (6.23) in Theorem 6.6.

Remark 7.2 The reader may wonder whether the corresponding Q, R need not imply good performance. However, this is not the case as the inner loop is specially designed

Table 7.2 List of constraints employed in the closed-loop simulation

Variables	Constraints
Maximum rotor speed $\omega_{\max}$	12.83 rpm
Maximum pitch angle $\theta_{\max}$	90°
Maximum pitch rate $\dot{\theta}_{\max}$	8 degrees per second
Minimum pitch angle $\theta_{\min}$	0°
Minimum pitch rate $\dot{\theta}_{\min}$	−8 degrees per second

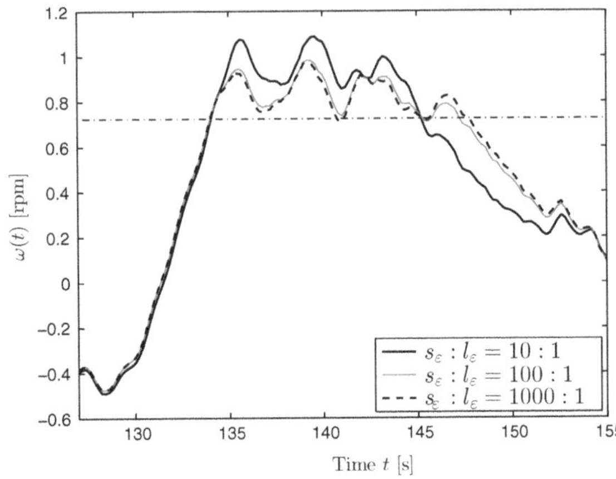

Fig. 7.3 Soft constraint on rotor speed deviation with different quadratic and linear weights, s_ε and l_ε, respectively, on the slack variables. Dash-dot line denotes the maximal speed deviation of 0.725 rpm

to meet frequency domain criteria in preference to time-domain criteria and hence implicitly the corresponding Q, R are appropriate.

Tuning of the weights of the quadratic penalty S_ε and l_1-norm penalty L_ε in the cost (7.14) is dependent on the trade-off between the duration and peak of the constraint violations. To illustrate, Fig. 7.3 shows simulation results of constraint violations on the rotor speed deviation, where the diagonal entries of S_ε are $s_\varepsilon > 0 \in \mathbb{R}$ and the elements of L_ε are $l_\varepsilon > 0 \in \mathbb{R}$. It is clearly seen in Fig. 7.3 that increasing the relative importance of the quadratic penalty s_ε compared to the l_1-norm penalty l_ε results in prolonging the duration of the constraint violation but a reduction in the peak violation. Given that over speeding the rotor beyond a certain threshold could potentially cause the turbine to trigger a temporary shut down, there is a good argument that it is more favourable to minimise the size of violation. Nonetheless, the l_ε need to be chosen large enough to ensure the soft constraint to be exact, which implies the slack variables are enforced to zero whenever a feasible solution of the optimisation is possible (Scokaert & Rawlings, 1999).

7.6 Numerical Results and Discussions

The aims of this section are to demonstrate the benefits gained by deploying the proposed MPC layer on top of the nominal robust feedback controller. The separate nature of the proposed design offers a transparent framework to distinguish the marginal improvement by deploying various features, for example, the capability of handling upcoming disturbance information or constraint violations, into the nominal controller. This provides insights for wind turbine manufacturers to evaluate the benefits against the associated cost of each feature. The results were obtained from closed-loop simulations upon a high-fidelity wind turbine.

In Sect. 7.6.1, time history samples were extracted from simulation results to investigate the performance gained by utilising upwind measurements of disturbance and constraint handling. Three constraint violation cases are investigated as follows:

1. Constraint violations on the rotor speed,
2. Constraint violations on the magnitude of the blade pitch angles,
3. Constraint violations on the blade pitch rates.

This is then followed by analysis of full results obtained from simulations under various wind conditions in Sect. 7.6.2.

7.6.1 Case Studies: Benefits of Utilising Upcoming Measurments and Constraint Handling

This section examines the improvement from deploying the MPC layer in cases when constraint violations are expected. Three controllers, detailed in Table 7.3, were compared: (i) the baseline nominal feedback-only controller (7.9), denoted as FB; (ii) a preview controller that utilises the advance measurements of rotor speed and blade disturbance but no constraint handling capability, denoted as FB/FF$_{\omega_d M_d}$; and (iii) the final controller is a preview and constraint-aware controller and its control law obtained by solving Algorithm 7.4.2 on-line, denoted FB/MPC$_{\omega_d M_d}$. Three types of constraints were employed in this chapter, thus, comparisons were made for three classes of constraint violations.

7.6.1.1 Constraint Violations on the Rotor Speed

It can be clearly seen in Figure 7.4 that both controllers, FB/FF$_{\omega_d M_d}$ and FB/MPC$_{\omega_d M_d}$, outperform the baseline controller FB since they use advance measurements. Furthermore, the time history of rotor speed in Fig. 7.4a indicates that the constraint-aware controller FB/MPC$_{\omega_d M_d}$ anticipated and avoided violating the maximum rotor speed constraint. Similar blade loads and pitch activities are observed in Fig. 7.4b, c, d.

Table 7.3 Various control configurations employed in this study

Controllers	Availability of of ω_d	Availability of $\tilde{M}_d$	Constraint handling
FB (baseline)			
FB/FF$_{\omega_d}$	✓		
FB/FF$_{\omega_d M_d}$	✓	✓	
FB/MPC$_{\omega_d}$	✓		✓
FB/MPC$_{\omega_d M_d}$	✓	✓	✓

With the nominal feedback controller as the baseline, additional features such as feed-forward knowledge and constraint handling are incrementally augmented into the feedback controller

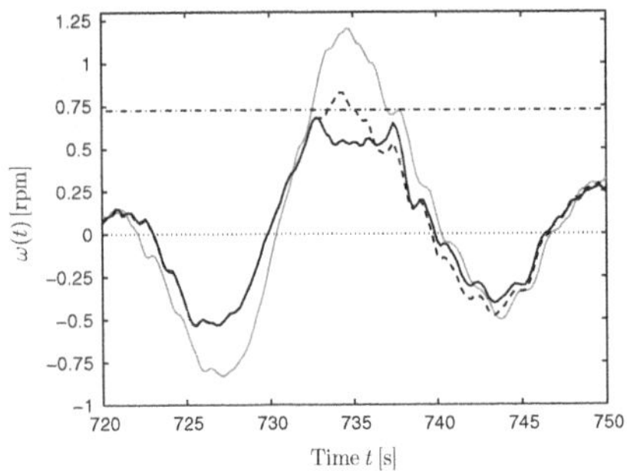

(A) Time history of the rotor speed deviation.

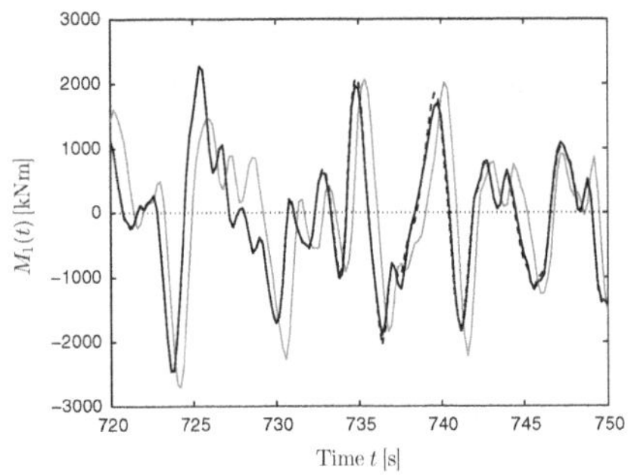

(B) Time history of the flap-wise blade root bending moment of blade 1.

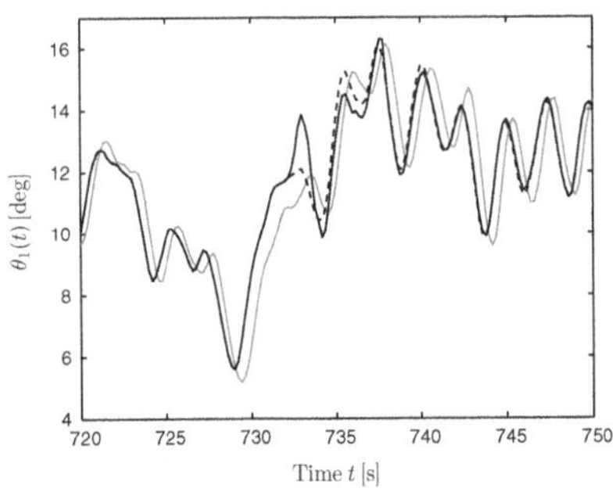

(C) Time history of the pitch angle of blade 1.

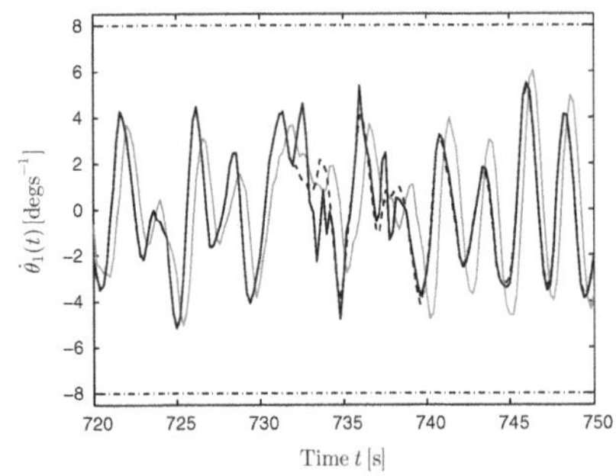

(D) Time history of the pitch rate of blade 1.

Fig. 7.4 Simulation results upon the NREL 5MW turbine operating in a wind case with the mean speed of 19 ms^{-1} and turbulence intensity of 14%, showing the performance of the various controllers studied in this chapter. Similar behaviours are obtained for the remaining blades. (i) Thin grey line: FB. (ii) Thick dash line: FB/FF$_{\omega_d M_d}$. (iii) Thick solid line: FB/MPC$_{\omega_d M_d}$. (iv) Dash-dot line: constraints. (v) Dot line: the performance targets

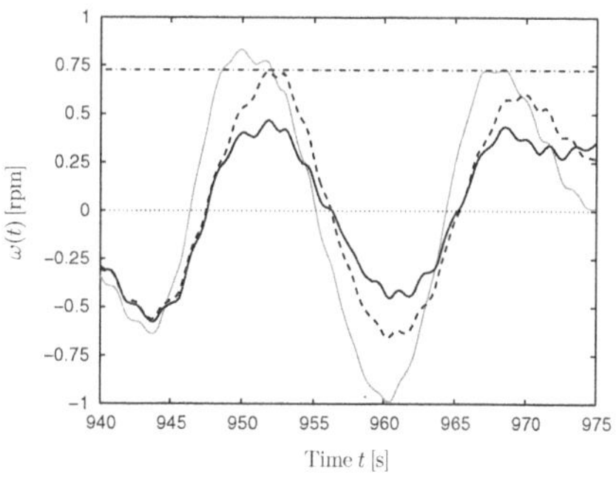

(A) Time history of the rotor speed deviation.

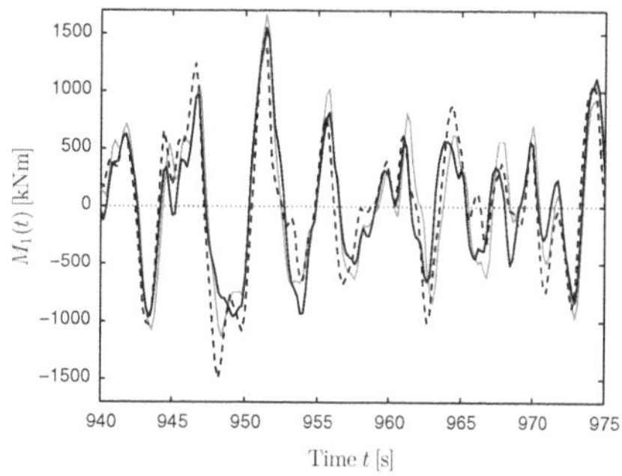

(B) Time history of the flap-wise blade root bending moment of blade 1.

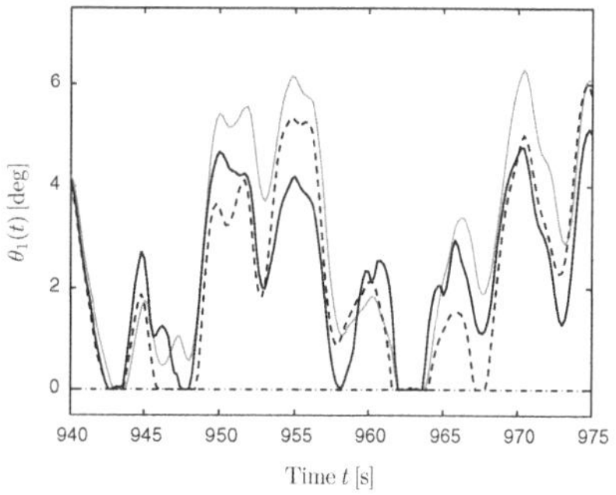

(C) Time history of the pitch angle of blade 1.

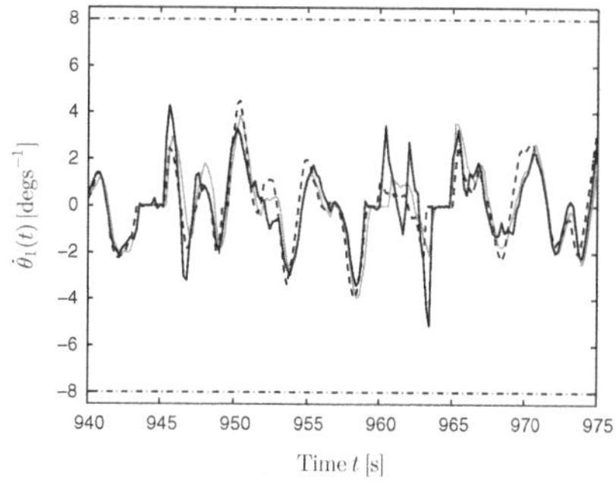

(D) Time history of the pitch rate of blade 1.

Fig. 7.5 Simulation results upon the NREL 5MW turbine operating in a wind case with the mean speed of 13 ms^{-1} and turbulence intensity of 14%, showing the performance of the various controllers studied in this chapter. Similar behaviours are obtained for the remaining blades. (i) Thin grey line: FB. (ii) Thick dash line: FB/$FF_{\omega_d M_d}$. (iii) Thick solid line: FB/$MPC_{\omega_d M_d}$. (iv) Dash-dot line: constraints. (v) Dot line: the performance targets

7.6.1.2 Constraint Violations on the Blade Pitch Angle

Figure 7.5 presents results where the blade pitch actuators steer near the lower limit. Owing to advance knowledge of the disturbance, it is not surprising that the preview controllers, FB/$FF_{\omega_d M_d}$ and FB/$MPC_{\omega_d M_d}$, perform better on rotor speed tracking and blade load reduction than the baseline controller FB as shown in Fig.7.5a, b. In addition, Fig. 7.5a, b reveals that significant reductions in the rotor speed deviation and flap-wise blade bending moments were achieved by the controller FB/$MPC_{\omega_d M_d}$ which is aware of the actuator constraints, as evident in Fig. 7.5c.

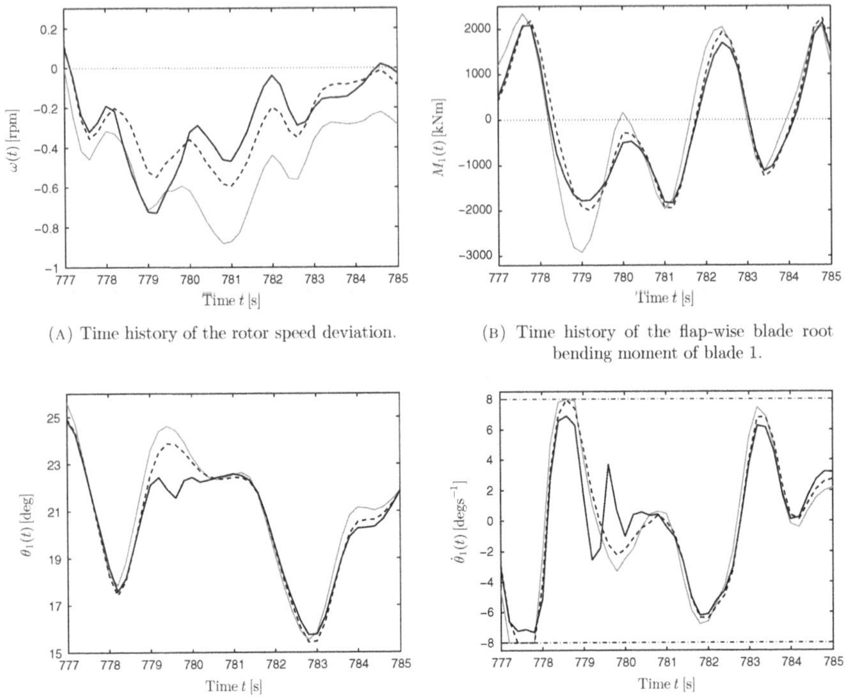

(A) Time history of the rotor speed deviation.

(B) Time history of the flap-wise blade root bending moment of blade 1.

(C) Time history of the pitch angle of blade 1.

(D) Time history of the pitch rate of blade 1.

Fig. 7.6 Simulation results upon the NREL 5MW turbine operating in a wind case with the mean speed of 23 ms^{-1} and turbulence intensity of 18%, showing the performance of the various controllers studied in this chapter. Similar behaviours are obtained for the remaining blades. (i) Thin grey line: FB. (ii) Thick dash line: FB/FF$_{\omega_d M_d}$. (iii) Thick solid line: FB/MPC$_{\omega_d M_d}$. (iv) Dash-dot line: constraints. (v) Dot line: the performance targets

7.6.1.3 Constraint Violations on the Blade Pitch Rate

The situation where the pitch actuators operate close to their maximum rate is illustrated in Fig. 7.6. As shown in Fig. 7.6b, better reductions in the flap-wise blade bending moment were yielded by the constraint-aware preview controller, FB/MPC$_{\omega_d M_d}$, that foresees the pitch rate violations, as indicated in Fig. 7.6d. In general, it is apparent that constraint anticipations of pitch rates provide the least benefit in contrary to pitch angle and rotor speed constraints. Nevertheless, this seems plausible because pitch actuators operate at the maximum rate for a relatively short period, typically less than one second. Thus, the controller that foresees the rate constraint might only achieve limited benefits given the fast blade dynamics.

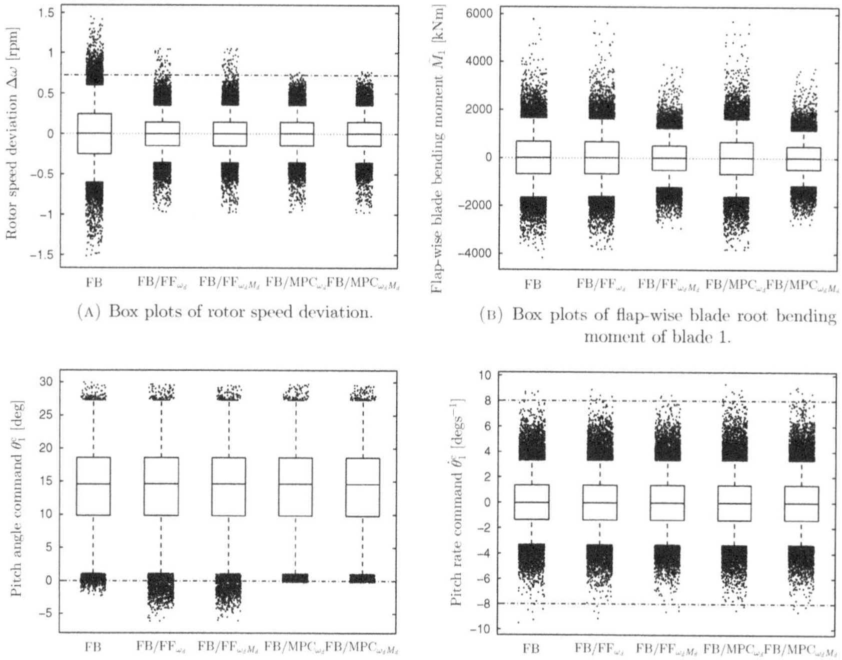

(A) Box plots of rotor speed deviation.

(B) Box plots of flap-wise blade root bending moment of blade 1.

(C) Box plots of pitch angle command of blade 1.

(D) Box plots of pitch rate command of blade 1.

Fig. 7.7 Simulation results upon the NREL 5MW turbine under various wind cases with mean wind speed ranging from 13 to 23 ms^{-1} and turbulence intensity of 14–18%, showing the performance comparison between the various controller configurations studied in this chapter. Similar results are obtained for the remaining blades. Dash-dot line represents the constraints and dot lines denote the performance targets

7.6.2 Simulation Results Under Various Wind Cases

This section presents results obtained from closed-loop simulations under numerous wind cases. These wind cases are characterised by a mean speed between 13 and 23 ms^{-1}, spanning a large range of above-rated wind conditions, and turbulence intensity ranging from 14 to 18%. Two more preview controllers, summarised in Table 7.3, were considered: FB/FF$_{\omega_d}$ and FB/MPC$_{\omega_d}$ and such controllers are the same as FB/FF$_{\omega_d M_d}$ and FB/MPC$_{\omega_d M_d}$, respectively, except that the upcoming measurements of blade disturbance are not available. The performance box plots of the result data generated from 180 sets of 20-min simulations are shown in Fig. 7.7. Each box represents the first and third quartiles whilst the band within the box represents the median of the dataset. The whiskers denote 5 and 95% quantiles. The data beyond the whiskers are considered as outliers, indicated by dots.

Figure 7.7a presents the box plot of the rotor speed performance. It can be clearly seen that the preview controllers achieved better reductions in rotor speed deviation compared to the baseline feedback-only controller, owing to the upcoming measurements of rotor speed disturbance trajectories. Moreover, the constraint-aware controllers manage to retain the rotor speed within the limit for most of the time despite the fact that the constraints on rotor speed were occasionally relaxed to ensure feasibility of the constrained optimisation of the MPC layer. These results indicate that proper management of constraint violations can lead to significant reductions in rotor speed.

Referring to Fig. 7.7b, the box plot shows the blade flap-wise root bending moment. Performance achieved by the controllers without the upcoming measurements of blade loads M_d was almost identical. In contrast, better reductions in the flap-wise blade moments were yielded by both preview controllers with knowledge of future blade loads. Close inspection between these two controllers reveals that the constraint-aware controller FB/MPC$_{\omega_d M_d}$ performed slightly better than the preview-only controller FB/FF$_{\omega_d M_d}$, which is consistent with the previous findings in Sect. 7.6.1.

The pitch angle and pitch rate command of blade 1 are illustrated by box plots in Fig. 7.7c, d, respectively. Note that the pitch angle command θ_1^c and pitch rate command $\dot{\theta}_1^c$ are considered instead of the pitch actuator activities because such activities are almost identical for the five alternative control strategies and investigation of pitch command signals reveals how constraints were handled by each controller. Results in Fig. 7.7c, d are as expected, that the MPC-based controllers were well aware of the constraints and managed the commands to avoid operating beyond those constraints. Few outliers beyond the pitch angle and rate constraints can be observed in Fig. 7.7c, d. This is conceivable since soft constraints are imposed on those variables.

The statistical properties of the data obtained from turbine simulation are summarised in Fig. 7.8 and Table 7.4. Figure 7.8a demonstrates the performance reductions achieved by the proposed controllers compared to the baseline controller in terms of standard deviations, whilst Fig. 7.8b shows how the proposed controllers performed in the maximum load cases. Finally, Fig. 7.8c reveals that the chance of constraint violations was reduced significantly by incorporating controllers with constraint handling capability. In general, the results suggest that controllers with more features perform better than those without them. Nevertheless, the cost and technical complexity associated with each layer are different, for example, in practice, the preview measurement of rotor disturbance could be estimated based on an averaged wind speed of few point measurements across the rotor disk whereas accurate estimations of the upcoming blade disturbance are less trivial. As a whole, these results could be used as a representative guide on the potential performance benefits achievable by additional control features.

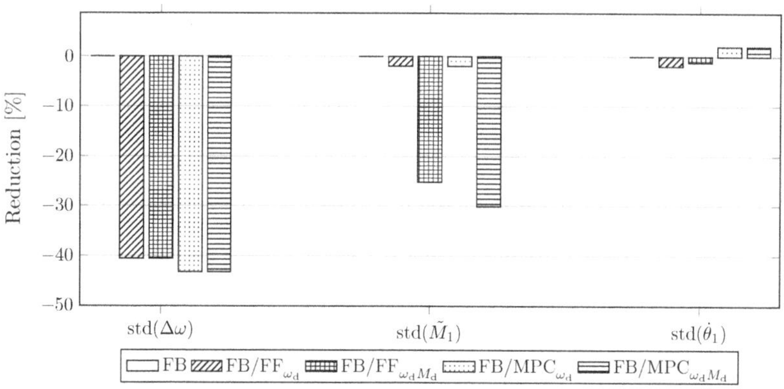

(A) Standard deviation of turbine loads from simulation results. Comparisons of reductions achieved by the proposed controllers compared to the baseline.

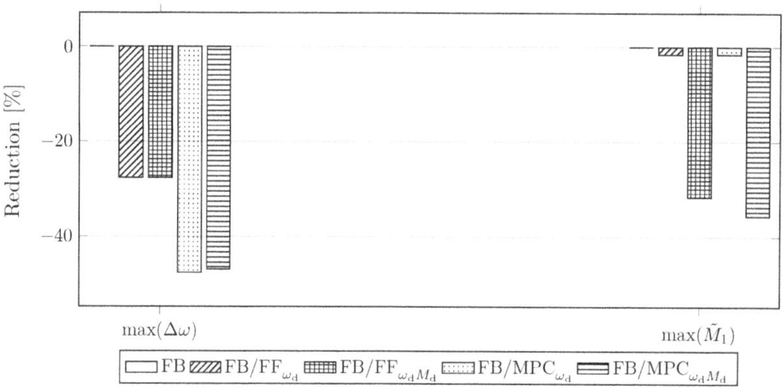

(B) Reductions on the extreme loads from simulation results.

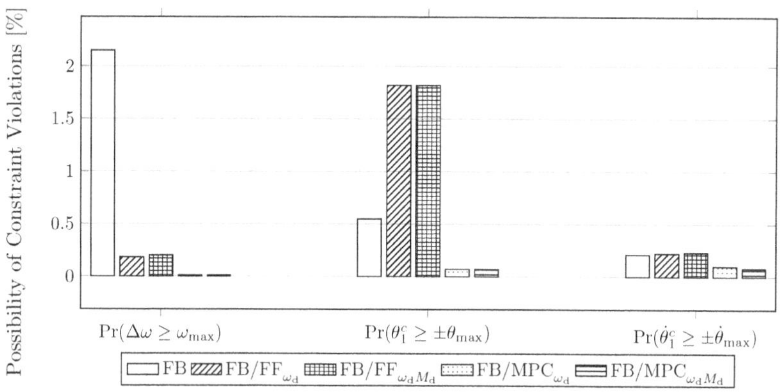

(C) Probability of constraint violations over the entire simulation period.

Fig. 7.8 Illustration of statistical properties of the data obtained from turbine simulations

Table 7.4 Summary of simulation results upon the NREL 5MW turbine

	FB (Baseline)	FB/FF$_{\omega_d}$	FB/FF$_{\omega_d M_d}$	FB/MPC$_{\omega_d}$	FB/MPC$_{\omega_d M_d}$
std($\Delta\omega$) [rpm]	0.37 (100.00%)	0.22 (59.46%)	0.22 (59.46%)	0.21 (56.76%)	0.21 (56.76%)
std($\tilde{M}_1$) [MNm]	1.03 (100.00%)	1.01 (99.03%)	0.77 (74.76%)	1.01 (98.06%)	0.72 (69.90%)
std($\dot{\theta}_1$) [degs^{-1}]	2.53 (100.00%)	2.48 (98.02%)	2.50 (98.81%)	2.58 (101.98%)	2.58 (101.98%)
max($\Delta\omega$) [rpm]	1.45 (100.00%)	1.05 (72.41%)	1.05 (72.41%)	0.76 (52.41%)	0.77 (53.10%)
max($\tilde{M}_1$) [MNm]	5.79 (100.00%)	5.70 (98.45%)	3.96 (67.36%)	5.70 (98.45%)	3.73 (64.42%)
Pr($\Delta\omega \geq \omega_{\max}$) [%]	2.15	0.18	0.20	0.01	0.01
Pr($\theta_1^c \geq \pm\theta_{\max}$) [%]	0.55	1.82	1.82	0.07	0.07
Pr($\dot{\theta}_1^c \geq \pm\dot{\theta}_{\max}$) [%]	0.21	0.22	0.23	0.10	0.08

Noted that std and max denote the standard deviation and maximum value, respectively. Pr represents the possibility of constraint violations. Differences in percentage to the baseline controller are represented in brackets

7.7 Summary

In this chapter, the modular predictive feed-forward layer based on top of a robust output-feedback blade-pitch compensator is presented. The separate nature of the controller structure enables clear and transparent performance comparisons and this was demonstrated by a comprehensive set of results obtained from closed-loop high fidelity turbine simulations upon a variety of different controllers. In particular, simulation scenarios of the controller constraint handling capabilities on the rotor speed and the pitch actuators were examined. The proposed control scheme incorporating the knowledge of the upcoming wind and constraints achieved remarkable 43 and 30% reductions in the rotor speed and flap-wise blade moment standard deviations. Additionally, the chance of constraint violations on the rotor speed were significantly down from 2.15 to 0.01%, compared to the baseline controller.

Nonetheless, the performances of the proposed controller were evaluated under assumptions of perfect knowledge of the upcoming wind. In some wind turbine designs, the baseline controllers could be output-feedback gain-scheduling controllers given the non-linearity of the blade-pitch control problem. This chapter only considered the predictive control layer formulation based on one single robust output-feedback compensator. Furthermore, the control design for below-rated wind conditions was not included in this chapter. In addition, the errors and uncertainties in wind measurements, for example, wind evolutions, were not fully considered in this study, and remain topics of future research.

References

Bossanyi, E., Savini, B., Iribas, M., Hau, M., Fischer, B., Schlipf, D., ... Carcangiu, C. E. (2012). Advanced controller research for multi-MW wind turbines in the UPWIND project. *Wind Energy, 15*(1), 119–145.

Bossanyi, E. A. (2003). Individual blade pitch control for load reduction. *Wind Energy, 6*(2), 119–128.

Dunne, F., Pao, L., Wright, A., Jonkman, B., Kelley, N., & Simley, E. (2011). Adding feedforward blade pitch control for load mitigation in wind turbines: Non-causal series expansion, preview control, and optimized FIR filter methods. *49th AIAA aerospace sciences meeting including the new horizons forum and aerospace exposition* (pp. 1–17). American Institute of Aeronautics and Astronautics: Reston, VA.

Dunne, F., Pao, L. Y., Wright, A. D., Jonkman, B., & Kelley, N. (2011). Adding feedforward blade pitch control to standard feedback controllers for load mitigation in wind turbines. *Mechatronics, 21*, 682–690.

Geyler, M., & Caselitz, P. (2008). Robust multivariable pitch control design for load reduction on large wind turbines. *Journal of Solar Energy Engineering, 130*(3), 031014.

Harris, M., Hand, M., & Wright, A. (2005). LIDAR for Turbine Control (Tech. Rep.).

Jonkman, B. (2009). *TurbSim User's Guide* (Tech. Rep.). National Renewable Energy Laboratory (NREL).

Jonkman, J., & Buhl Jr., M. (2005). *FAST User's Guide* (Tech. Rep.). National Renewable Energy Laboratory (NREL).

Jonkman, J., Butterfield, S., Musial, W., & Scott, G. (2009). *Definition of a 5-MW Reference Wind Turbine for Offshore System Development* (Tech. Rep.). Golden, CO: National Renewable Energy Laboratory (NREL).

Kodama, N., Matsuzaka, T., Tuchiya, K., & Arinaga, S. (1999). Power variation control of a wind generator by using feed forward control. *Renewable Energy, 16*, 847–850.

Laks, J., Pao, L., Simley, E., Wright, A., Kelley, N., & Jonkman, B. (2011). Model predictive control using preview measurements from LIDAR. In *49th AIAA*. Reston, VA: American Institute of Aeronautics and Astronautics.

Laks, J., Pao, L., Wright, A., Kelley, N., & Jonkman, B. (2011). The use of preview wind measurements for blade pitch control. *Mechatronics, 21*(4), 668–681.

Leithead, W., Neilson, V., & Dominguez, S. (2009). Alleviation of unbalanced rotor loads by single blade controllers. In *European Wind Energy Conference*.

Lio, W. H., Jones, B. L., Lu, Q., & Rossiter, J. A. (2015). Fundamental performance similarities between individual pitch control strategies for wind turbines. *International Journal of Control*, 1–16.

Lio, W. H., Rossiter, J. A., & Jones, B. L. (2014). A review on applications of model predictive control to wind turbines. *2014 UKACC International Conference on Control (Control)* (pp. 673–678). IEEE: Loughborough.

Lu, Q., Bowyer, R., & Jones, B. (2015). Analysis and design of Coleman transformbased individual pitch controllers for wind-turbine load reduction. *Wind Energy, 18*(8), 1451–1468.

Raach, S., Schlipf, D., Haizmann, F., & Cheng, P. W. (2014). Three dimensional dynamic model based wind field reconstruction from LIDAR data. In *The science of making torque from wind*.

Schlipf, D., Fischer, T., & Carcangiu, C. (2010). Load analysis of look-ahead collective pitch control using LIDAR. In *Proceedings of 10th German Wind Energy Conference*.

Schlipf, D., Schuler, S., Grau, P., & Martin, K. (2010). Look-ahead cyclic pitch control using LIDAR. In *The science of making torque from wind*.

Scokaert, P. O. M., & Rawlings, J. B. (1999). Feasibility issues in linear model predictive control. *AIChE Journal, 45*(8), 1649–1659.

Selvam, K., Kanev, S., van Wingerden, J. W., van Engelen, T., & Verhaegen, M. (2009). Feedback-feedforward individual pitch control for wind turbine load reduction. *International Journal of Robust and Nonlinear Control, 19*(1), 72–91.

Spencer, M. D., Stol, K. A., Unsworth, C. P., Cater, J. E., & Norris, S. E. (2013). Model predictive control of a wind turbine using short-term wind field predictions. *Wind Energy*, *16*(3), 417–434.

Towers, P., & Jones, B. L. (2016). Real-time wind field reconstruction from LIDAR measurements using a dynamic wind model and state estimation. *Wind Energy*, *19*(1), 133–150.

Wang, N., Johnson, K. E., Wright, A. D., & Wright, A. D. (2012). FX-RLS-based feedforward control for LIDAR-enabled wind turbine load mitigation. *IEEE Transactions on Control Systems Technology*, *20*(5), 1212–1222.

Zhang, Y., Cheng, M., & Chen, Z. (2013). Proportional resonant individual pitch control for mitigation of wind turbines loads. *IET Renewable Power Generation*, *7*(3), 191–200.

Chapter 8
Conclusions and Future Work

This final chapter is organised as follows: Sect. 8.1 presents a summary of the original contributions in this thesis. This is followed by a final conclusion including discussions of the limitations within this thesis in Sect. 8.2. Finally, recommendations for future work are discussed in Sect. 8.3.

8.1 Summary of Original Contributions

The original contributions of the thesis can be summarised as follows:

1. The links between various IPC designs upon the rotating and non-rotating reference frames, namely the Coleman transform-based IPC (e.g. Bossanyi, 2003), Single-blade IPC (e.g. Leithead, Neilson, & Dominguez, 2009 and Clarke transform-based IPC (e.g. Zhang, Cheng, & Chen, 2013), were studied. The robust stability margin was used to prove that these IPC design methods are fundamentally equivalent. Furthermore, the load spectra of the blade loads, obtained from simulation results upon an NREL 5MW turbine, verified that the performance of these different IPC designs is also similar.
2. The interactions between the blades and tower upon rotating and non-rotating coordinate frames were investigated and based on these insights, the observability of the modelled system was established. Subsequently, an estimator was then formulated to reconstruct the fore-aft velocity of the tower-top from the blade load measurements. Simulations were conducted to demonstrate the performance of the estimator. In addition, a control law was also designed to show that the estimated tower signals can be used for tower load attenuations.
3. The problem of formulating a feed-forward MPC design on an existing output-feedback controller was illustrated, where the feed-forward MPC would potentially introduce an additional feedback loop that alters the original closed-loop

W. H. (Alan) Lio, *Blade-Pitch Control for Wind Turbine Load Reductions*,
Springer Theses, https://doi.org/10.1007/978-3-319-75532-8_8

dynamics. A modular MPC layer design was then proposed, with a view to retaining the closed-loop robustness and frequency-domain properties of the given output-feedback controller. Key conditions were derived to prove that the proposed MPC layer for handling the advance knowledge impacts upon the existing closed-loop system if and only if constraint violations are expected.

4. The proposed MPC layer design upon the existing blade pitch controller was implemented to a non-linear simulation wind turbine. To overcome the infeasibility issues under the realistic scenario, soft constraints were employed and the tunings of the penalties on the slack variables were particularly investigated. In addition, the benefits of the proposed MPC layers were clearly demonstrated upon the high-fidelity simulation turbine, where significant improvements on turbine structural load reductions were shown. Moreover, the separate nature of the proposed control structure revealed clearly the performance improvement gained by utilising the measurement of the upcoming wind and constraint handling capabilities, compared to the baseline feedback controller.

8.2 Final Conclusions

The aims of the thesis were to investigate the blade-pitch control in wind turbine load reductions. By achieving these objectives, the contributions of this thesis have a number of implications to the research community and industry.

Firstly, in Chap. 4, the proof of the equivalence between different IPC design methods could imply that choice of IPC technique rests largely on the preferences and implementation. Secondly, in Chap. 5, the tower vibration estimation and control design could offer a means to justify the cost of using some highly reliable but expensive blade load sensors and also obviate the need for tower-top sensor redundancy. Thirdly, in Chap. 6, a modular MPC layer design allows a systematic incorporation of future measurements of disturbance and constraint handling capabilities into a given output feedback design, which can be synthesised by robust frequency-domain methods or an existing in-house industrial design. Lastly, in Chap. 7, the separate nature of the proposed MPC layer design enables clear and transparent quantifications of the benefits gained from utilising measurements of the upcoming wind and constraint knowledge.

Nonetheless, there are some limitations to the analysis and design proposed within this thesis. For example, in Chap. 4, the equivalence of the IPC design is established based on an assumption that the rotor speed is constant. Another example is in Chaps. 6 and 7, the analysis of the robustness properties of the existing closed-loop is limited to situations where constraints were expected not to be active. Since constraint satisfactions are less crucial in wind turbines where instability caused by failure of constraint fulfilments is less likely, thus, robust constraint requirements are not particularly considered in this thesis.

As a final remark, this thesis investigated the load reduction control in wind turbines via individual blade pitching and use of measurements of the upcoming wind. In addition, some clarifications and systematic design guidelines were provided on these these topics. In the near future, more new actuators and sensors will be available to wind turbines, especially to offshore floating turbines, many innovative load reduction solutions are anticipated to come.

8.3 Recommendations for Future Work

Possible directions for the future work identified in this thesis are listed as follows:

1. Some of the loads on the turbine blades and rotor are deterministic to some degree, for example, the tower shadow effect and wind shear. Given this repetitive nature, iterative learning control (ILC) could be effective in rejecting such loads, where the load signals from the previous period can be used to update the control law over the next period. Some studies (e.g. Houtzager, van Wingerden, & Verhaegen, 2013; Tutty, Blackwell, Rogers, & Sandberg, 2014) demonstrated the efficacy of ILC in wind turbines, however, the performance of the ILC could be degraded by the presence of variations of the rotational period of the rotor. Therefore, a new design, that takes into account the variations of rotor speed, is worthy of further study.
2. As the turbine size increases, many actuators on the turbine rotor become available, for example trailing edge flaps (Barlas & van Kuik, 2010). These trailing edge flaps on the blade can be adjusted independently in response to flap-wise blade moments for attenuating unsteady loads on the turbine blades. Similar to IPC techniques, some of the existing controller designs employ Coleman transformations. Therefore, the analysis of performance similarities of IPC designs from Chap. 4 can possibly extend to the trailing edge flap control design.
3. Most of the IPC designs aim to attenuate unsteady blade loads in the above-rated wind conditions, whilst in the below-rated wind conditions, the blade pitch are kept at zero angles. Thus, there is an opportunity to utilise the blade pitch control for attenuating loads in the below-rated wind conditions as long as the power generation remains unaffected.
4. In this thesis, measurements of the upcoming wind are assumed to be known a priori perfectly. The errors and uncertainties in wind measurement are not fully considered in the current work. Nonetheless, the recent development of wind estimation technology enables fast and accurate wind field reconstruction. This reconstructed wind field can be employed in the feed-forward control and such studies remain topics of the future work.
5. As mentioned in Chap. 3, the application of some recently developed MPC theories to wind turbines is still in its infancy, such as tube-based robust or stochastic MPC. In standard MPC, constraint violations might occur since there are mismatches between the predictions and the actual dynamical behaviour of the plant. To lower the number of violations, constraints are often tightened based on a priori knowledge about the uncertainties and disturbance, resulting in a slow and con-

servative design. A fast data-driven strategy is proposed by a recent study (Carrau, Liniger, Zhang, & Lygeros, 2016), where the constraint tightening parameter is tuned on-line based on the discrepancy between the model and the plant. Future work could look to implementing this on-line tuning method to the MPC layer to improve the rate of constraint satisfactions on the rotor speed.

6. An increasing number of wind turbines begin installing in a remotely offshore site since the wind conditions at the sea are less turbulent and it is more economical to do so. Floating platforms, that are employed by turbines in the deep sea, introduce extra dynamics to the turbines, which makes blade pitch control design more challenging. Model predictive control in floating wind turbines is worth investigating given that the algorithm could handle the constraint requirements of the floating platform by blade pitching.

References

Barlas, T., & van Kuik, G. (2010). Review of state of the art in smart rotor control research for wind turbines. *Progress in Aerospace Sciences*, *46*(1), 1–27.

Bossanyi, E. A. (2003). Individual blade pitch control for load reduction. *Wind Energy*, *6*(2), 119–128.

Carrau, V., Liniger, A., Zhang, X., & Lygeros, J. (2016). Efficient Implementation of Randomized MPC for Miniature Race Cars. *2016 European Control Conference.*

Houtzager, I., van Wingerden, J. W., & Verhaegen, M. (2013). Wind turbine load reduction by rejecting the periodic load disturbances. *Wind Energy*, *16*(2), 235–256.

Leithead, W., Neilson, V., & Dominguez, S. (2009). Alleviation of unbalanced rotor loads by single blade controllers. In *European Wind Energy Conference.*

Tutty, O., Blackwell, M., Rogers, E., & Sandberg, R. (2014). Iterative learning control for improved aerodynamic load performance of wind turbines with smart rotors. *IEEE Transactions on Control Systems Technology*, *22*(3), 967–979.

Zhang, Y., Cheng, M., & Chen, Z. (2013). Proportional resonant individual pitch control for mitigation of wind turbines loads. *IET Renewable Power Generation*, *7*(3), 191–200.